With AI Towards Sustainable Building Structures

Diego Apellániz
Berlin University of the Arts
Dr.-Ing. (doctoral graduate)

CRC Press
Taylor & Francis Group
Boca Raton London New York

CRC Press is an imprint of the
Taylor & Francis Group, an **informa** business
A SCIENCE PUBLISHERS BOOK

Cover credit: The image used on the cover is created by the author.

First edition published 2026
by CRC Press
2385 NW Executive Center Drive, Suite 320, Boca Raton FL 33431

and by CRC Press
4 Park Square, Milton Park, Abingdon, Oxon, OX14 4RN

CRC Press is an imprint of Taylor & Francis Group, LLC

Library of Congress Cataloging-in-Publication Data (applied for)

ISBN: 978-1-032-82064-4 (hbk)
ISBN: 978-1-032-82805-3 (pbk)
ISBN: 978-1-003-50632-4 (ebk)

DOI: 10.1201/9781003506324

Typeset in Palatino Linotype
by Prime Publishing Services

Preface

The evolution of artificial intelligence (AI) is perhaps the most notable technological innovation in recent years. The influence of AI on daily life is evident: people frequently turn to chatbots for guidance or information, much of today's advertising uses AI-generated images and videos, and AI recommends music and shows we might like based on our past preferences. AI is also revolutionizing industry and science, although making a meaningful impact requires researchers and practitioners to have both a thorough grasp of AI and substantial expertise in the specific area. Based on this premise and my over ten years of professional experience as a structural engineer, self-taught programmer, and researcher, this book represents my modest contribution to leveraging AI in the fight against climate change by enabling sustainable practices and minimizing the building sector's environmental footprint. Although the book concentrates on building structures due to my background, these elements are central to buildings and account for a large share of their environmental impact. Additionally, the chapters of the book are crafted in straightforward language, allowing both professionals across the building industry—not just structural engineers—and the general public to follow, comprehend, and hopefully enjoy them.

The term "artificial intelligence" was coined in 1956 to refer broadly to systems designed to mimic human intelligence. Over time, this field has progressed to today's advanced neural networks, which frequently outperform humans in many respects. This book does not aim to provide an academic history of AI's evolution, as fascinating as that may be, but instead focuses on the state of the art methods with the greatest potential to benefit the building industry. It is also important to note that there is not a single "AI", but rather a family of very different methods that must be assessed independently. Furthermore, my aim with this book is not to make mere bold predictions about AI's future, but to explain the fundamentals of these methodologies and provide an overview of the building industry and its carbon footprint. This foundation empowers readers to form their own judgments on the proposed approaches, guidelines, and reflections in this book for the sustainable application of AI in construction. The scope of the book ranges from established AI applications often integrated into building projects to emerging developments whose full potential for sustainable building projects we can only begin to glimpse. Throughout the book, any software, platforms, or services mentioned in relation to AI applications are provided solely as illustrative examples. Their inclusion does not imply that they are the best in the market, the only available options, or that I have any commercial agreement or affiliation with their developers. Finally, the ethics of AI, both in general and with a particular focus on the building sector, are addressed in the closing chapter.

Most AI fields evolve so rapidly that keeping pace with the latest developments can prove challenging. Although I have endeavored to include the most recent advances, certain sections, particularly those describing specific digital workflows in building projects and those related to specific versions of AI models, may become somewhat outdated over time. Nevertheless, the fundamental principles presented here will remain valid, helping readers to understand ongoing progress and to assess how future AI innovations might be applied to enhance the sustainability of building projects. This rapid evolution of AI mirrors the mid-nineteenth century, when pioneering engineers were discovering how to apply new materials like cast and wrought iron and the first steels to bridge construction. As the eminent engineer and locomotive designer Robert Stephenson famously remarked about a later bridge by one of his peers, "If Brunel's bridge stands, mine"—referring to his Britannia Bridge with its wrought iron tubular design—"was a waste of iron." In the same way, AI is advancing toward ever greater power and energy efficiency, and it is ultimately up to every stakeholder in the building industry to leverage this technology for the greatest positive impact.

Throughout my career as a structural engineer, I have been driven by a curiosity to explore how digital tools, bespoke software, and algorithms can enhance engineering tasks. Initially, I turned to coding to automate routine work, freeing me to focus on the creative aspects of structural design. I soon realized that developing efficient digital workflows and writing code presented an engineering challenge as engaging as creating resilient, functional, and elegant structures. To share these workflows and tools, I launched a YouTube channel, which has now reached four thousand subscribers. Some of the tools I have created over the years are open-source applications used by thousands of practitioners worldwide. This enthusiasm led to my Doctor of Philosophy (PhD) research on applying parametric design and machine learning to integrate sustainability assessments in the early stages of building design. Yet I wanted to make these insights accessible not only to researchers but also to the general public, despite how rewarding the academic journey was. Moreover, the latest advances in generative AI emerging after my thesis warranted further investigation into their potential sustainable applications in the building sector. With this book, I have channeled my drive and expertise into a work I hope readers will find informative, insightful, and engaging.

Contents

List of Abbreviations

1D	One-dimensional
2D	Two-dimensional
3D	Three-dimensional
ACP	Panama Canal Authority
AGI	Artificial General Intelligence
AI	Artificial intelligence
API	Application Programming Interface
AR	Augmented reality
ASCE	American Society of Civil Engineers
ASR	Automatic speech recognition
AWS	Amazon Web Services
Ax	Adaptive Experimentation
BIM	Building Information Modeling
BREEAM	Building Research Establishment Environmental Assessment Method
BSc	Bachelor of Science
CAD	Computer-aided design
CAG	Cache augmented generation
CASP	Critical Assessment of protein Structure Prediction
CFD	Computational fluid dynamics
CLIP	Contrastive Language–Image Pretraining
CLT	Cross-laminated timber
CNN	Convolutional neural network
CO2	Carbon dioxide
CO2e	Carbon dioxide equivalent
COVID-19	Coronavirus disease 2019
CPU	Central processing unit
CUA	Computer-using agent
DGNB	German Sustainable Building Council
DQN	Deep Q-Network
EC3	Embodied Carbon in Construction Calculator
EJ	Exajoule
EPD	Environmental Product Declaration
ERP	Enterprise Resource Planning

EU	European Union
EUCB-D	Embodied Carbon of European Buildings Database
FEM	Finite Element Method
GAN	Generative adversarial network
GDT_TS	Global Distance Test – Total Score
GPS	Global Positioning System
GPT	Generative Pre-trained Transformer
GPU	Graphics processing unit
GRPO	Grouped Relative Policy Optimization
Gt	Gigaton
GUPC	Grupo Unidos por el Canal
HPC	High-performance computing
HVAC	Heating, Ventilation and Air Conditioning
IFC	Industry Foundation Classes
IoT	Internet of Things
IP-Adapter	Image Prompt adapter
IT	Information technology
LCA	Life-cycle assessment
LED	Light-emitting diode
LEED	Leadership in Energy and Environmental Design
LiDAR	Light Detection and Ranging
LLM	Large language model
LoRA	Low-Rank Adaptation
LSTM	Long Short-Term Memory
LVL	Laminated veneer lumber
MCP	Model Context Protocol
MCTS	Monte Carlo Tree Search
MDP	Markov decision process
MEP	Mechanical, Electrical and Plumbing
ML	Machine learning
MSc	Master of Science
MuJoCo	Multi-Joint dynamics with Contact
MVP	Minimum viable product
OCR	Optical Character Recognition
OOP	Object-oriented programming
PC	Personal computer
PhD	Doctor of Philosophy
PINNs	Physics-Informed Neural Networks
PPO	Proximal Policy Optimization
R&D	Research and development
RAG	Retrieval-augmented generation
RGB	Red, Green, Blue
RLHF	Reinforcement learning from human feedback
SaaS	Software as a Service

SDS	Score Distillation Sampling
SHM	Structural health monitoring
SLAM	Simultaneous Localization and Mapping
SVM	Support vector machine
TPU	Tensor Processing Unit
UI	User interface
U-Net	U-shaped convolutional neural network
US	United States
VAE	Variational Autoencoder
ViT	Vision Transformer
VLM	Vision-language model
VR	Virtual reality
VRAM	Video Random Access Memory

AI and Building Industry Impacts in Perspective

1

I would venture to say that there is a unanimous consensus that climate change is undesirable: the debate merely centers on whether to tax or ban contributing sectors, invest in sustainable technologies, or let the free market self-regulate. Among the contributing sectors to our climate challenges, the building industry stands out as a major, often overlooked player. This chapter is the starting point of an investigative journey to explore how cutting-edge AI techniques, which turn out to have a significantly smaller environmental footprint than the construction industry, could potentially transform it, towards a more sustainable future, starting with its core: the building structures.

Environmental Costs of Construction

"Buildings always have a profound impact on our mood and well-being; they shape our perceptions and can uplift our daily lives" (de Botton 2008). This statement highlights the essential role buildings have in human civilization. They offer physical protection, a sense of safety, and spaces where individuals can reside, work, and flourish. On a personal level, we often spend most of our time in buildings—be it a home, an office, a school, or a community center. They safeguard us from the elements, facilitate social and cultural interactions, and provide a private space for personal well-being. From a broader perspective, buildings are central to economic growth and social unity. They generate employment opportunities during their construction and in the functions they support, such as businesses, educational institutions, or healthcare facilities. Additionally, buildings influence the character of cities, towns, and neighborhoods, mirroring cultural values and local customs. Iconic architecture can also become a

DOI: 10.1201/9781003506324-1

symbol of regional identity, attracting visitors and contributing to cultural heritage and tourism.

Given this profound influence of buildings on our lives and communities, it is crucial to consider their long-term impacts on our planet. The original definition of sustainability defines this concept as "meeting the needs of the present without compromising the ability of future generations to meet their own needs" (Burton 1987), capturing economic, social and environmental implications. Despite the potential positive economic and social impacts of buildings just mentioned, the construction, operation, and eventual demolition of buildings are associated with significant environmental trade-offs. The production of building materials such as concrete, steel, and glass requires substantial amounts of energy and natural resources, often leading to considerable carbon emissions and environmental degradation. These emissions are then exacerbated during the manufacturing process. Once constructed, buildings continue to impact the environment through their operational phase. They are consumers of energy, primarily for heating, cooling, lighting, and running various electronic devices and appliances. This energy consumption contributes to global energy demand, influencing the rate of fossil fuel depletion and increasing greenhouse gas emissions. Additionally, the demolition of buildings generates significant waste and releases stored carbon dioxide, further contributing to global warming. The built environment also affects local ecosystems and water resources through factors such as increased surface runoff, reduced biodiversity, and the strain on local water supplies.

Globally, the built environment sector accounted for approximately 34% of total final energy consumption, equating to around 150 exajoules (EJ) out of 442 EJ in 2022 (IEA 2023), the most recent year for which data is available. During the same period, carbon dioxide (CO_2) emissions from building operations and construction escalated to nearly 10 gigatons (Gt), which represented 37% of total global CO_2 emissions (United Nations Environment Programme and Global Alliance for Buildings and Construction 2024). The significant environmental impact of the building sector might surprise many, especially since prominent discussions and media, such as Al Gore's 2006 documentary "An Inconvenient Truth," often emphasize other sectors like automotive and transportation, overlooking the substantial contributions of buildings to global emissions.

Another unexpected major impact on the global level from the building sector may be the enormous resource consumption, especially in the area of mineral material. Sand is crucial in construction, used primarily as an aggregate in concrete, mortar, and plaster. Even though there is plenty of sand in oceans and deserts, suitable construction sand is difficult to find because it requires specific angular grains free from impurities, unlike the smooth and often salt-contaminated grains found in desert and sea sands.

Therefore, it typically comes from riverbeds or quarries. The severity of this problem is reflected in the increasingly widespread and expensive, but now necessary technology of crushing stones to produce aggregates (Drew et al. 2002). Additionally, this scarcity has led to the emergence of mafias in some developing countries, who profit from controlling the extraction and distribution of this valuable construction material (Mahadevan 2019).

Challenges in Sustainable Constructions

Since ancient times, buildings have been required to fulfill three basic functions: strength, utility, and beauty. Or "Firmitas, Utilitas, Venustas," as the Roman architect and engineer Vitruvius stated in his treatise "De Architectura." A building may also be sustainable, but not at the expense of any of the former qualities. For instance, no building developer would want to own a sustainable office or residential building that potential tenants would find impractical, ugly, or unstable. They would most certainly find themselves bankrupt if they neglected any of Vitruvius' principles.

What exactly is a sustainable building? Focusing on the environmental aspect of sustainability, it refers to a structure that has a minimal environmental impact throughout its life-cycle. Regarding building materials, there are established strategies to achieve this (United Nations Environment Programme and Yale Center for Ecosystems + Architecture 2023): minimizing waste and constructing with fewer resources by prioritizing resource-efficient designs, favoring renovations over demolitions and new builds, and using recycled and reused materials.[1] Another approach is to shift to bio-based materials like timber or bamboo, which have low greenhouse gas emissions during production and can capture CO_2 as new trees grow in their place. Additionally, improving conventional materials and processes is key, such as using low-carbon cement in concrete or employing advanced technologies for producing recycled steel. For the building's operation, a sustainable structure incorporates energy-efficient technologies and systems, such as high-efficiency Heating, Ventilation, and Air Conditioning (HVAC) systems and light-emitting diode (LED) lighting, to significantly cut energy consumption. It may also generate its own energy through solar panels, wind turbines, or geothermal systems. Water efficiency is enhanced by using low-flow fixtures and water recycling systems, all aimed at reducing water waste.

[1] Reused materials are directly repurposed in their existing form for the same or new functions, while recycled materials are processed and transformed into new products.

However, if sustainable buildings provide such significant benefits for our environment, why aren't all new constructions designed to meet these standards? One primary reason is the economic viability of building projects. The initial costs associated with adopting eco-friendly designs can be intimidating for developers. On one side, the demand for "green premiums" (Grainger 2024) on rents for buildings has been increasing steadily in recent years, particularly in Europe, North America, and Asia.[2] However, the current shortage of residential buildings might be diverting some focus from environmental issues, as the pressing need for housing overshadows long-term sustainability goals. Beyond market dynamics, governments also provide subsidies for eco-friendly building projects.[3] Another strategy for governments and regulatory bodies to promote sustainable construction economically is to impose taxes on building projects that fail to meet certain sustainability standards.[4] The downside to these regulatory measures is the burdensome paperwork that developers and building owners must complete to comply with these programs.

Despite these economic and regulatory efforts to encourage sustainable building, the reality of implementing such practices is fraught with complexity. Renovating existing buildings or using eco-friendly materials often leads to complex design scenarios not fully addressed by current building standards, requiring careful evaluation. To obtain green building certification—whether for attracting tenants, accessing green financing, or publicity—extensive documentation must be submitted to certification bodies, often necessitating the involvement of an external consultancy. The design team must stay informed about the latest eco-friendly materials and clean technologies, as well as their market availability. They must also be equipped to conduct preliminary analyses efficiently, presenting building owners with various design options that weigh sustainability against other factors like cost, construction complexity, and user comfort, enabling informed decision-making. An interdisciplinary collaboration among planning disciplines—such as architecture; structural engineering; Mechanical, Electrical and Plumbing (MEP) and building physics—is

---

[2] This trend is largely driven by heightened environmental awareness among consumers and tenants, who are increasingly alarmed by the approach to the critical threshold of 1.5 degrees Celsius of global warming (IPCC 2021).

[3] This is not just an altruistic approach, but it has been proven that the costs of climate change induced damages to infrastructure, property, agriculture, and human health cost the world several million $ per hour, so in fact, these incentives are rather an investment (Bennett 2023).

[4] One of these policies would be the so-called carbon-pricing, which through emissions trading systems and carbon taxes, aims to internalize the environmental costs of greenhouse gas emissions, effectively making greener building practices not only more ethically appealing but also economically advantageous.

essential for conducting these comprehensive analyses, which will soon be mandatory in several countries, making it crucial for design offices to adapt,[5] regardless of the owner's sustainability goals. Lastly, during construction, acquiring eco-friendly materials can be challenging due to rising demand and limited supply. Additionally, finding suitable existing buildings for extensions and renovations, along with available stocks of reusable building elements, can be difficult due to insufficient data.

AI and the Environment: Tools and Trade-offs

AI is on track to be the most impactful technology of the 21st century. The pioneers of AI were awarded the Nobel Prize in 2024.[6] New advancements in AI models are emerging weekly. AI applications top the download charts on both Android and Apple smartphone markets. AI adoption is expanding rapidly across various industries. Investments from governments and venture capital firms into AI projects are growing exponentially. Regulating AI has become a top priority for political leaders worldwide.

However, despite constant exposure to AI applications and content, only 51% of adults acknowledge recognizing which products and services utilize AI (Ipsos 2023), let alone understanding how these AI algorithms function. The upcoming chapters of the book will introduce these algorithms at a basic level to illustrate their applicability in the building industry, but essentially, most current AI applications employ a subset of machine learning known as deep learning. Deep learning algorithms are based on neural networks, which mimic the structure of the human brain and consist of interconnected layers of nodes or "neurons." They are designed to detect patterns in input data without needing explicit instructions. At its core, deep learning enables machines to process and analyze large amounts of data in a manner that resembles human thought processes, leading to more intuitive and intelligent system behavior.

[5] The Buildings Breakthrough announced at COP28 United Nations Climate Change Conference in December 2023 aims to make low emission, climate-resilient buildings the new norm by 2030 and so far, it has the support of 27 countries, the European Union (EU) commission and 18 international initiatives. In the coming years, companies making little to no progress on tackling their carbon footprints including across their real estate portfolios will have nowhere to hide because the documentation of the building carbon footprint will be mandatory.

[6] John Hopfield and Geoffrey Hinton were awarded the 2024 Nobel Prize in Physics for their foundational work in artificial neural networks, while Demis Hassabis and John Jumper won the Nobel Prize in Chemistry for their contributions to protein structure prediction with AI.

The infrastructure required to run AI applications starts with the hardware. In recent years graphics processing units (GPUs) have become a cornerstone for AI workloads. These chips were originally created for rendering graphics in videogames and other visual applications, using extensive parallel computing to rapidly process the multitude of repetitive calculations required for each pixel. AI leverages this concept by utilizing GPUs' parallel computing power to efficiently handle the large-scale, repetitive calculations happening within neural networks. For instance, each neuron in a layer is generally independent of others when it comes to mathematical computations, allowing them to be processed in parallel. This synergy between graphics rendering and AI will be further explored in subsequent chapters. It has been exploited by GPU manufacturers such as NVIDIA, which have seen their market capitalization multiply since the advent of AI (Clark 2024).

Although GPUs can be installed into personal computers to develop and test small AI models like simple image classifiers, current state-of-the-art large language models (LLMs) and other large generative AI models require the aggregated power of multiple graphics processing units (GPUs) and even the more AI oriented Tensor Processing Units (TPUs) housed in servers within data centers. This is where High-performance computing (HPC) comes into play. By distributing the workload across numerous GPUs and TPUs, HPC systems can significantly shorten the time required for processing data and training the AI, enhancing the efficiency and scalability of deep learning operations. Housing an HPC cluster requires physical floor space to hold and support the weight of racks of servers, which typically contain not only GPUs and TPUs but also a variety of other critical components.

It takes considerable power to operate and cool these data centers. It's crucial to differentiate between the resources used during the training phase and those used during the inference (or usage) phase.[7] Training an AI model typically demands significantly greater computational resources and, by extension, energy consumption, than querying an already-trained model, although the latter can still accumulate sizable energy usage when scaled to large numbers of requests. Interestingly enough, the so-called

[7] Training a neural network involves adjusting its internal parameters (weights) based on data input to learn specific patterns or functions. This process is computationally intensive as it requires repeatedly processing large datasets and adjusting the network's weights via backpropagation to minimize errors. Running a neural network, also known as inference, occurs after the network is trained. It involves inputting new data into the trained network to obtain predictions based on the learned patterns. Running a network is generally less computationally demanding than training, as it only involves forward propagation through the network's pre-set weights.

open-source generative AI models[8] are typically trained by a corporation or institution using the computing resources of a data center and then released to the public, who can run these models locally on their personal computers or dedicated servers. Nevertheless, most AI operations take place at data centers and will continue to do so for the foreseeable future (Wevolver 2024).

There are numerous challenges in computing the global energy demand and environmental impact of AI. Data centers perform many functions other than AI like web hosting, streaming services, cloud computing and much more. In HPC environments within data centers, the dynamic and shared use of resources across these various applications makes it challenging to precisely attribute specific amounts of power and resources to AI tasks. Furthermore, the corporations responsible for the largest AI models are not prone to disclose the energy consumption of their AI operations for competitive reasons. Nevertheless, recent studies (Sandalow et al. 2024) estimate that AI operational emissions are less than 1% of total global CO_2 emissions and the emissions related to manufacturing and maintaining AI hardware and infrastructure may be equal to the operational emissions—so another 1%. Therefore, if we compare current environmental impact of the building industry, as presented in a previous section, and that of AI, it would be 37% of total global CO_2 emissions produced by the former vs. a maximum 2% of total global CO_2 emissions by the latter. There are of course other environmental indicators beyond CO_2 emissions and the future projected emissions of AI are uncertain.[9] This comparison helps to establish a reasoned argument for the potential application of AI to help mitigate the environmental impact of the building industry, particularly in the construction of structures.

Potential for AI in Normalizing Sustainable Building Structures

"The fact that CO_2 is transparent and odorless is the greatest tragedy for humankind" is one of the architect and structural engineer Prof. Werner Sobek's statement on sustainability. Since this greenhouse gas is invisible

[8] Open-source generative AI models are publicly available AI systems whose code and often parameters are made freely accessible, allowing anyone to use, modify, and improve them for a wide range of creative or practical applications.

[9] On the one hand, the demand for AI services is expected to grow at a significant rate next years (Morgan Stanley 2024). On the other hand, the energy efficiency of data centers has improved (Robles y Zepf and Schaefer 2024) and there is a trend within generative AI models towards not larger but more efficient model architectures that require less energy demand than the largest models released so far.

and undetectable to our senses, people often fail to recognize the danger it poses until its consequences are already upon us—by which time meaningful mitigation can be far more difficult. Humans need to perceive something with their senses to truly grasp its presence and impact.

For this reason, the focus on building structures is so important. From all elements and processes that make up a building, when we talk about building structures, we are referring to the tangible "bones" or physical framework of a building. Because structures are visible and can often be experienced directly (whether through their shape, layout, or materials), people naturally grasp and appreciate improvements made to them. This physicality leads to public awareness and acceptance: When a sustainable change is evident in the shape of cross-laminated timber beams, a high-efficiency frame, or a newly retrofitted building skeleton, it commands attention and sparks curiosity.

This book does not intend to undervalue the significance of operational impacts or the impacts of other elements and processes that, while not as apparent as a building structure, also deserve exploration. However, building structures are often the largest contributors to upfront carbon emissions, as the next chapter will elaborate. These emissions occur during the brief period before the building becomes operational and addressing them is crucial for making an immediate positive impact on global warming and preventing its irreversible effects.

Before embarking on exploring AI's potential in sustainable building structures, it's essential to first discuss, in Chapter 2, the use of life-cycle assessment as a metric for measuring sustainability, specifically focusing on the carbon footprint of building structures. This metric is necessary to ensure that AI applications genuinely contribute positively and aren't just superficial green washing. Additionally, as a conclusion to the introduction, Chapter 3 examines the history of digitalization in the design and construction of building structures, which forms the foundation for AI applications.

Following this introduction, the subsequent chapters of the book will explore how the different AI technologies can transform the construction industry into a more sustainable field. The text will not only spotlight the areas where this transformation is currently underway but also address the challenges faced in expanding AI application more broadly.

Measuring the Sustainability of Building Structures

2

First, we need to understand how to measure improvements in the sustainability of building structures before diving into the discussion on how different AI technologies can be applied to this task. Otherwise, there is a significant risk, as in many other industries, that AI is not adding any additional value but is merely being used to capture the public's attention. Worse yet, AI might be used to greenwash high-carbon building projects by shifting focus away from the carbon hotspots. This chapter outlines an established methodology for measuring the environmental impact of buildings, details the main approaches in reducing the carbon footprint of building structures, and discusses the primary challenges of implementing these in real projects, drawing on my decade-long career as a structural engineer. Not only does this chapter lay the groundwork for sensibly applying AI to enhance building sustainability, but it also provides the knowledge to distinguish between genuinely sustainable buildings and those that are not.

How to Measure the Carbon Footprint

"How Much Does Your Building Weigh, Mr. Foster?"[1] This provocative question was posed by Buckminster Fuller, a visionary architect and

[1] The question 'How Much Does Your Building Weigh, Mr. Foster?' is not only a provocative inquiry by Buckminster Fuller but also the title of a documentary film about the architect Norman Foster. Released in 2010, the film explores Foster's career and his pioneering approach to modern architecture, which emphasizes sustainability and technological innovation. This question encapsulates the film's exploration of how Foster's design philosophy is influenced by considerations of material efficiency and environmental impact, reflecting the themes that have defined his contributions to architecture.

DOI: 10.1201/9781003506324-2

pioneer in the field of lightweight structures, to Norman Foster, a prominent British architect. The question came up during their discussion about Foster's design of the Sainsbury Centre for Visual Arts[2] in the 1970s, aimed at prompting him to reflect on the efficiency of his building. This question spurred Foster's team into action, prompting them to analyze the designs of all building components to compute the total mass, which amounted to 5,328 tons. However, the major revelation was not the mass of the visible filigree steel structure but the hidden concrete foundation that constituted most of the weight. Although the term "sustainability" had not yet been coined, this introspective analysis did more than just calculate the total mass of the structure—a goal Fuller likely had in mind to benchmark against typical values for steel roofs and assess the efficiency of Foster's design. It also involved identifying the most significant contributors to the building's mass. This early exploration into what we now recognize as sustainability analysis was undoubtedly a precursor to the types of sustainability assessments carried out in the design phases of building projects today. However, as highlighted in the previous chapter, striving for material efficiency is just one of several strategies to achieve sustainable buildings. How can materials like steel be compared to bio-based materials such as timber? Is building mass the only metric available? What are the benefits of refurbishing buildings compared to constructing new ones? Additionally, how can we calculate the (in)famous carbon footprint of a building and attempt to offset it through other activities?

The tool employed nowadays to answer these questions is the life-cycle assessment (LCA). LCA is a methodology "for assessing environmental impacts associated with all the stages of the life-cycle of a commercial product, process, or service," which can also be applied to evaluate the environmental impact of a building, whether it is already constructed or still in the design phase. The first step in calculating the life-cycle assessment of a building is to compute the material quantities or masses of the building, similar to what Norman Foster's team did for the Sainsbury Centre for Visual Arts. The second step involves accessing a database that details the environmental related to each building material. These databases, compiled by public entities, manufacturers and life-cycle assessment software distributors, include Environmental Product Declarations (EPDs). EPDs are official documents that quantify the environmental impacts of a single building product or material throughout its life-cycle, providing the necessary data to multiply against the material quantities calculated in the first step. Finally, the carbon emissions produced during the operational phase of the building, as well as those from various construction site processes independent of the building

[2] Readers are highly encouraged to look up all referenced buildings in this book.

materials, must also be computed. The fact that a building comprises numerous products and processes that need quantifying, combined with the availability of tens of thousands of EPDs related to buildings, makes the task of calculating the life-cycle assessment of a building to determine its environmental impact throughout its entire life cycle a particularly tedious process. This complexity is schematically represented in Figure 1. How this calculation can be highly automated to utilize its results for reducing the environmental impacts of a building from the early design stages will be explored in the upcoming chapters.

The results of the life-cycle assessment are expressed through various environmental indicators that quantify the environmental impacts of the building. These include Global Warming Potential, Ozone Depletion Potential, Water Footprint, among others. Although these terms may sound technical, it is the Global Warming Potential, expressed in terms of carbon dioxide equivalents (CO_2e),[3] that is commonly referred to as the carbon footprint of the building. In fact, overwhelmed by the various indicators provided by life-cycle assessments, many designers opt to focus primarily on the building's carbon footprint to simplify the decision-making process. This focus is due to the direct relationship between the carbon footprint and global warming, and the urgent need to address global warming,

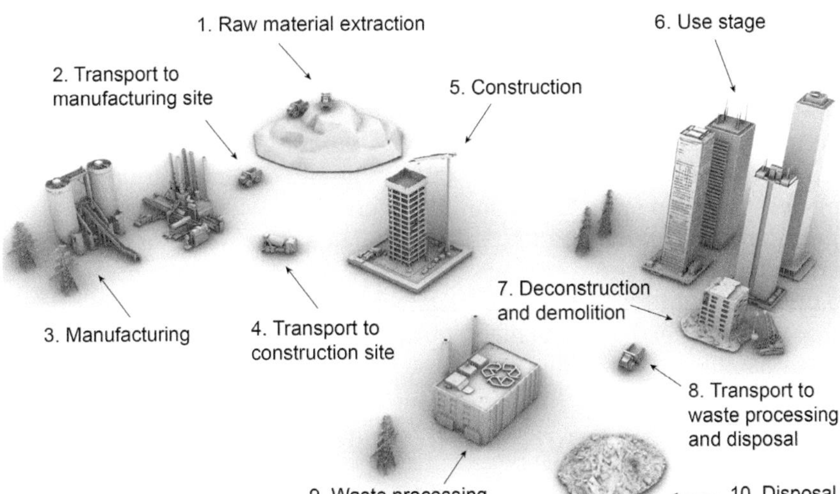

Figure 1. Life-cycle stages of a building.

[3] CO_2e is a standard unit for measuring carbon footprints, expressing the impact of each different greenhouse gas in terms of the amount of CO_2 that would create the same amount of global warming. This unit is used to aggregate the impact of different greenhouse gases, allowing them to be compared and summed into a single figure.

as explained in the first chapter. According to recent studies (Apellániz et al. 2024), decision-makers primarily use life-cycle assessment results to compare different preliminary designs, identify the so-called carbon hotspots in terms of materials or building elements like slabs, columns, and walls, and to benchmark the carbon footprint of the design against standard values. In this way, the life-cycle assessment becomes a tool to reduce the building's carbon footprint during the design phase.

Needless to say, building owners who invest resources in such studies to reduce the carbon footprint expect benefits beyond just environmental impact. One simple approach is to publicize the reduced carbon footprint by comparing it to common activities understood by the general public, such as equating the reduction to the carbon footprint of hundreds of oceanic flights. Another option is to position the building as carbon-neutral. This could involve offsetting the carbon footprint by planting enough trees to sequester the equivalent amount of carbon over the building's lifecycle or installing solar panels that not only provide renewable energy but also return enough power to the grid to compensate for the building's emissions. Increasingly, building owners and real estate developers are seeking green building certifications from organizations like Leadership in Energy and Environmental Design (LEED), Building Research Establishment Environmental Assessment Method (BREEAM), or German Sustainable Building Council (DGNB). These certifications, which can range from LEED Platinum to Gold, usually require a thorough life-cycle assessment as part of the application process. Applicants must optimize the building design across various environmental indicators while also addressing occupant health, comfort, and efficient operating systems. Credits may also be awarded for using recycled materials and designing for reusability of building elements. Green building certifications are becoming more prevalent and are valued by potential tenants as markers of sustainable and environmentally conscious buildings. While these certifications are a powerful tool for identifying sustainable buildings, it could be argued that they still need to refine their approach to more holistically capture the sustainability of a building.[4]

The Building Structure and Its Carbon Footprint

The building structure is the skeleton of the building and it plays a critical role in ensuring stability by transferring both vertical and horizontal loads

[4] Roudman (2025) continues the discussion around the environmental efficacy of the Bank of America Tower, focusing on the ongoing debate concerning its LEED certification and real-world performance versus initial green claims. Roudman explores various viewpoints, highlighting the discrepancies between the building's designed intentions and its operational outcomes.

into the ground. Vertical loads come from the building's own weight, including non-structural elements like partition walls, floor finishes, facades, and the weight of the occupants. Horizontal loads may arise from forces like wind or earthquakes. Beyond stability, the structure must also deliver user comfort by limiting deflections and vibrations caused by these external forces. Additionally, it is essential for the structure to maintain its durability throughout at least the entire lifecycle of the building.

A building structure is composed of horizontal slabs, which provide support for occupants, furniture, and equipment. These slabs connect all vertical elements, stabilizing them and transmitting wind forces to the lateral-force resisting walls. Certain types of slabs are supported by beams or truss elements that span between building columns. Both columns and walls bear the vertical loads from the slabs and transfer them down to the foundation. Columns typically handle the vertical loads exclusively and are straightforward to integrate into floorplans. Walls, in addition to bearing vertical loads, also manage lateral loads due to their stiffer cross sections. They are commonly integrated around elevator and staircase shafts or used as dividing elements to hinder fire propagation in the event of a fire. In basement levels, walls also serve as earth-retaining structures. The foundation, designed based on the magnitude of the building loads and the consistency of the ground, can either be a shallow foundation, such as isolated footings or foundation slabs, or a deep foundation consisting of piles.

As architect Louis Sullivan famously said, "Form follows function." In building structures, this principle means that beyond adhering to the laws of physics, the structure must be practical and accommodate the building's intended function. Structural engineers thus design the structure based on the expected use of the building, optimizing structural members to economically withstand expected forces. Figure 2 illustrates a mixed-use building where the left section houses a concert hall and the right is reserved for exhibitions. Inside the concert hall, columns are avoided to prevent obstructing audience views, necessitating trusses to support the wide-spanning slab. Conversely, the exhibition area, requiring large rooms but not the expansive column-free space of the concert hall, can utilize standard beams to support the slabs. However, these beams may need openings or perforations to accommodate building installations running beneath the slabs. Typically, building structures are covered to protect against environmental conditions, enhance energy efficiency, and for aesthetic reasons. Yet, in some cases, architects may leave structural elements like concrete or timber exposed as a deliberate aesthetic choice. When exposed, these elements must still meet stringent insulation and fire protection standards—particularly timber—to ensure they maintain their integrity and functionality over time. The design of buildings requires a high degree of collaboration across all planning disciplines—including

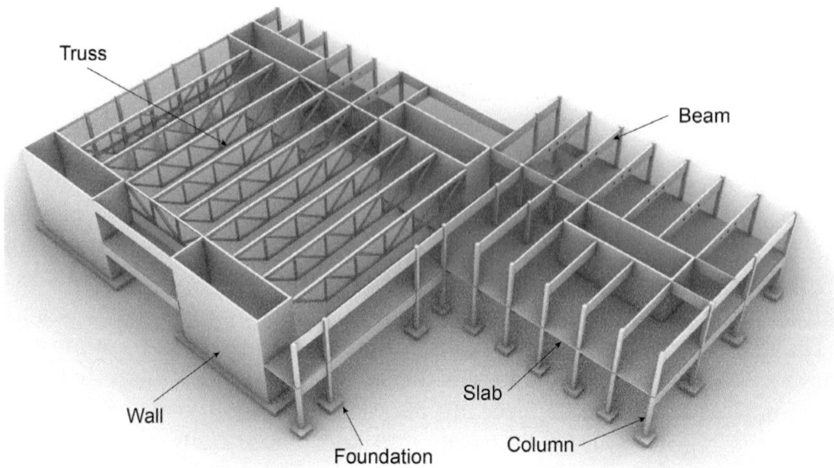

Figure 2. Elements of the building structure. Visualization with the roof slab removed for clarity.

architecture, structure, MEP, building physics, fire protection, and more—underscoring the complexity of integrating form and function in modern building design.

Interestingly, while carbon emissions from the operation of buildings—due to heating, electricity, and other factors—have traditionally been the largest contributors to their carbon footprint, recent studies[5] on modern multi-story buildings adhering to energy-efficiency standards indicate that embodied carbon has become equally significant. Embodied carbon refers to the emissions associated with the life-cycle of building materials, from the extraction of raw resources to their manufacturing processes. For example, the production of concrete, a fundamental component of most building structures, involves a calcination chemical reaction to generate cement clinker, which emits a substantial amount of CO_2. This process also includes emissions from the energy required to reach the necessary

[5] The publication "Net-zero buildings: Where do we stand?" (World Business Council for Sustainable Development and Arup 2021) is a comprehensive report that discusses the current status and future trajectories of net-zero carbon buildings. It examines six case studies of various building projects to illustrate the challenges and opportunities in achieving net-zero carbon emissions. The report emphasizes the critical need for whole life carbon assessments in building projects to understand and mitigate their carbon footprint throughout their lifecycle, from construction to operation and end-of-life stages. It advocates for a collaborative approach across the building industry to set clear, actionable targets and to share data openly, which would help in scaling up efforts towards significant global emissions reduction.

chemical reaction temperatures in the kiln, approximately 1450°C (2642°F). Moreover, the building structure itself accounts for about 50% to 60% of these initial emissions from the extraction and manufacturing phases. Further studies (Braune et al. 2021, Watson 2020) suggest that within the building structure, slabs are the primary contributors to this footprint, making them critical targets for optimization, followed by walls and foundations.

As mentioned in Chapter 1, some strategies to reduce the carbon footprint of building structures include minimizing waste and using fewer materials through designs that prioritize efficiency, shifting to bio-based materials that act as carbon sinks, favoring renovations over demolitions and new constructions, employing recycled and reused materials, and rethinking the production of traditional materials. Additionally, it's crucial to consider how these approaches can be integrated. This introduction to building structures sets the stage to explore these topics further before examining how AI can assist in these and other approaches toward more sustainable building structures.

Bringing Back Material-Efficient Structural Designs

Comparing current construction practices to those of the post-war era, when scarcity of resources, including building materials, was a critical issue, reveals significant potential for more efficient material use in building structures. During those times, engineers and builders developed innovative, material-saving typologies such as vaulted floors, shell roofs, and ribbed slabs. These designs utilized materials more efficiently to accommodate forces and transfer them to the foundation than the currently predominant flat slabs. Back then, designs were meticulously calibrated to adjust the amount of steel reinforcement in concrete down to the last cm^2, rather than rounding up quantities across large areas for simplicity. Additionally, there was a necessity to salvage building materials from ruined structures instead of relying solely on newly manufactured materials.

The major paradigm shift in construction practices is largely due to the evolution of construction costs. Today, labor costs are significantly higher compared to the cost of building materials, which has led to a preference for simpler shapes and designs that enable faster execution times. For instance, while flat slabs require relatively high quantities of reinforced concrete, the formworks for these homogeneous slabs are significantly simpler and faster to set up, and the concrete is easier to pour. This results in quicker construction times and lower labor costs. However, even though this design approach minimizes project costs, the use of materials - and consequently the embodied carbon—is far from efficient. This is one of the key reasons why embodied carbon has historically increased and now contributes as significantly as operational carbon in modern building designs.

There are various approaches in reintroducing material efficiency into structural designs within the current economic context, though each presents unique challenges. Material-efficient prefabricated solutions, such as precast concrete ribbed slabs or corrugated metal decking, offer benefits but can limit architectural creativity due to their standardized and typically rectangular shapes. Beyond prefabrication, several initiatives are attempting to revive traditional material-efficient shapes, combining them with state-of-the-art construction methods. However, these contemporary designs can introduce new challenges, including potentially higher construction costs, noise insulation issues in lighter floors, or the need for enhanced fire protection for steel rods that manage tension in arches or vaults. Despite these challenges, some research institutions are actively exploring solutions to these issues.[6]

It may be challenging for the untrained eye to identify efficient structural designs from those that are not. However, a useful rule of thumb can be drawn from the examples mentioned: if the structure is filigree, meaning that you can visually trace how forces are channeled from the slabs to the foundation in a direct, rather than a complex, Calatrava-esque manner, it is likely an optimized design. Prominent examples of material-efficient structural slab systems include Guastavino's vaulted slabs and Nervi's ribbed slabs. Fazlur Khan's innovative structural systems, which resist lateral wind forces in iconic Chicago skyscrapers like the John Hancock Tower, also exemplify significant efficiency. For those familiar with these constructions, an interesting takeaway is that striving for material savings can lead to elegant structural solutions. This is a core tenet of what is known as "structural art.[7] Such elegance is often more apparent in bridges than in buildings, as most buildings conceal their structures with floor and wall finishes. Nonetheless, engineers should continue to strive for designs that are both efficient and aesthetically pleasing. As William Baker, the engineer responsible for designing the Burj Khalifa, currently the world's tallest building, eloquently stated: "An engineer should aim for a design so beautiful that an architect is sad to cover it."

---

[6] The Form Finding Lab at Princeton and the Block Research Group at ETH Zurich are some of the institutions researching ways to reintegrate efficient structural shapes into modern construction, aiming to improve how these designs are utilized in building projects today.

[7] Structural art is a term that refers to the artful integration of form, efficiency, and economy in building structures. It was extensively defined and explored by Princeton's Professor David P. Billington in his influential book 'The Tower and the Bridge' (Billington 1983). In this work, Billington presents structural art not just as a technical achievement but as a creative discipline that elevates engineering to an art form, focusing on designs that meet practical needs while also being aesthetically pleasing.

Bio-Based Materials as Carbon Sinks and Other Alternative Materials

The role of timber and other bio-based building materials in acting as carbon sinks is well-documented (van der Lugt 2012) and illustrated in Figure 3. Through photosynthesis, trees absorb CO_2 and solar energy to produce wood, while simultaneously releasing oxygen. Remarkably, up to 0.9 tons of CO_2 can be stored in one cubic meter of wood. Utilizing bio-based materials as carbon sinks necessitates sustainably managed forests, where harvesting is limited to mature trees that grow and absorb carbon at a slower rate. The carbon absorbed during the trees' growth is then encapsulated in timber products used in building structures for the duration of their use. Harvesting strategies focus on fully-grown trees, with reforestation being crucial to replace the trees that are cut down. Timber's life cycle in building structures typically exceeds the natural lifespan of the source trees. With continuous regeneration in sustainably managed forests, a cycle of carbon absorption and storage is maintained. This enhances the carbon pool as new trees grow and replace those that were harvested, while the carbon remains sequestered in the existing timber structures.

Despite earlier representations of timber buildings as mere log houses, there has been significant technological advancement in the field of mass timber. Mass timber refers to a category of engineered wood that combines multiple layers of wood to form large, solid panels or beams. These materials, including cross-laminated timber (CLT), glued laminated timber (glulam), and laminated veneer lumber (LVL), offer

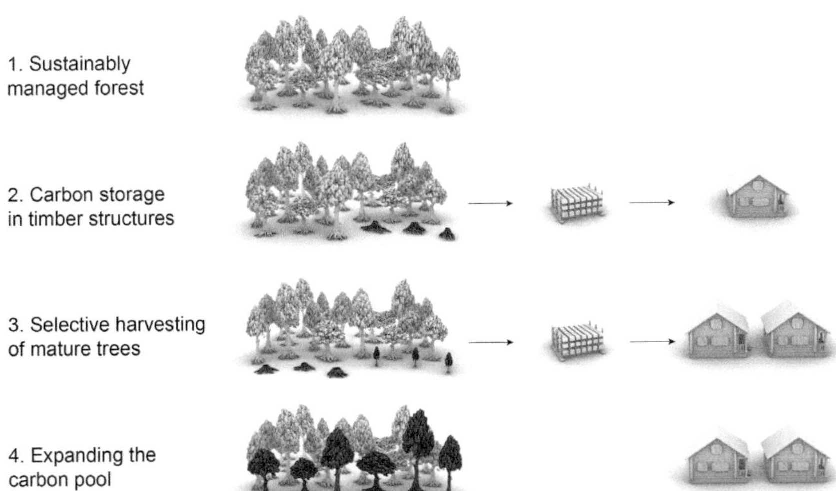

Figure 3. Sustainable forestry and carbon storage in timber structures.

exceptional strength and stability. The fire-resistance, durability against humidity and climate, and noise insulation properties of mass timber have been intensively studied, leading to evolved building regulations that now support the use of mass timber in multi-story constructions.[8] While the geometry of timber designs may be slightly constrained by the regular shapes of prefabricated timber elements, the appeal of timber for its sustainable properties, as well as its aesthetic and tactile qualities, is driving an increase in its architectural use. However, the supply of mass timber currently fails to meet this rising demand, resulting in higher construction costs. Consequently, many buildings initially planned with timber unfortunately end up being constructed with traditional materials such as reinforced concrete or steel.

Bamboo-based materials hold significant potential for use in construction, functioning as carbon sinks similar to timber. However, these materials require further research to fully understand their properties and potential applications. Additionally, building regulations must evolve to standardize their use in multi-story constructions. Alongside bamboo, other alternative materials such as rammed earth and clay masonry are beginning to be industrialized. While these materials do not directly act as carbon sinks, their adoption can significantly reduce the carbon footprint of buildings by replacing more carbon-intensive traditional materials. This shift contributes positively to sustainable construction practices and promotes environmental stewardship.

Building Renovations and When Not to Build

Building renovations are arguably the most effective approach for reducing the carbon footprint of existing structures, though they can also be the most challenging. Renovation involves significant alterations and improvements to a building or its components, typically aimed at enhancing functionality, aesthetics, or performance. This might include substantial modifications to the existing structure, major changes to the building's layout,[9] or significant upgrades to systems that boost energy efficiency and ensure compliance with modern standards.

The "Steglitzer Kreisel," a high-rise building in Berlin, Germany, serves as a representative example of the complexity of a renovation project.

[8] An example of a project utilizing bio-based materials is the Innovation Park Artificial Intelligence "IPAI" Campus. According to the project's description, "Most of the buildings take simple rectangular forms and have consistent heights of 27 metres, making them efficient to construct with modular grids and bio-based materials" (MVRDV 2023).

[9] The building layout refers to the arrangement of rooms and spaces within a building, as well as how these spaces are accessed and interconnected.

Previously an office building owned by the administration and vacated due to the presence of asbestos,[10] the new owner acquired it with plans to convert it into residential units (Wikipedia 2025). Demolishing the building was not an option due to an underlying subway line that could not be disturbed. Fortunately, the load requirements for residential use are lower than those for office spaces, allowing for the renovation without the need for intensive structural strengthening. However, a detailed assessment of the building's existing capacity,relying on project documents from the 1970s—was necessary. Additionally, local reinforcement measures were required to accommodate the new load distribution caused by the planned balconies for the residential units (Apellániz et al. 2019).

Although building renovations require meticulous planning - often more so than new building projects—and a detailed investigation of remaining structural capacity, they can significantly reduce carbon emissions, typically by 50–75% compared to equivalent new constructions (United Nations Environment Programme and Yale Center for Ecosystems + Architecture 2023). A case in point is the UP! Berlin building in Berlin, Germany. This former shopping mall was transformed into a modern office building. The renovation involved making strategic cut-outs in the original box-shaped structure to create a more dynamic appearance and enhance natural sunlight in the future office spaces. Removed areas of the building were then repurposed as an additional lightweight construction floor atop the existing structure, minimizing impact. This renovation project resulted in saving up to 10,000 tons of CO_2 equivalent, compared to what a new construction would have emitted (Gengnagel 2022). This amount of saved CO_2 is equivalent to the emissions from 10,000 passenger flights from London to New York, or approximately 3,300 people not driving their private cars for a year.

Recycling, Reusing and Rethinking Traditional Building Materials

Recycling involves breaking down and reprocessing materials that have reached the end of their useful life in a structure so they can be used in creating new building products. Steel is perhaps the most widely recycled building material. Recycling steel consumes significantly less energy

[10] Asbestos is a group of naturally occurring fibrous minerals that were commonly used in building materials for insulation and fireproofing due to their durability and resistance to heat. However, asbestos can be extremely hazardous when disturbed, as its fibers become airborne and pose serious health risks if inhaled, including lung diseases and cancer, which necessitates the complete evacuation of a building during its removal to ensure safety and compliance with health regulations.

compared to producing it from iron ore. It is estimated that recycling steel requires only about 30% of the energy needed to produce steel from raw materials, which makes it not only economically viable but also results in far fewer carbon emissions than manufacturing new steel. Additionally, the reduced carbon footprint of recycled steel is well documented in official EPDs. While recycling concrete as aggregates for newly poured concrete is common in applications like road construction, its use in building construction remains very limited.

Reusing materials, which involves utilizing them in their original form for the same or different purposes without significant processing, is unfortunately not yet a widespread practice. However, significant progress is being made in the area of "design for disassembly." This approach involves designing buildings and products in such a way that they can be easily dismantled at the end of their useful life, facilitating the reuse, recycling, and refurbishment of components. Green building certifications often grant credit points for this method. Design for disassembly is particularly feasible in steel structures with bolted connections, though the interfaces with reinforced concrete elements require careful consideration. A recurrent example is a parking building in a new urban development that is designed to be converted into an office building after 10-20 years, once the necessary permissions are granted. In such cases, every two stories of the parking building could be removed to achieve the required clear heights for the new office building, instead of demolishing the entire structure and starting from scratch. The removed parts would ideally be reused in another project. Close collaboration with demolition contractors is nevertheless essential to ensure they adhere to the planned disassembly instructions. An example of a failed execution of this approach is the cooling tower of the Schmehausen nuclear power plant in Germany (Schlaich Bergermann Partner 2025). Although the structural engineer, Jörg Schlaich, had designed the tower with a detachable cable net structure in 1974, intending for easy disassembly and cable reuse (notice the sustainable approach to the design even though the project dates from 1974!), the demolition contractor ultimately chose to demolish the tower with explosives in 1991.

The manufacturing of traditional building materials is also being rethought and redesigned to become more sustainable. This involves focusing on reducing environmental impacts, improving efficiency, and integrating renewable resources. In steel manufacturing, sustainability efforts include the use of electric arc furnaces, carbon capture technologies, and hydrogen as a cleaner reducing agent. Concrete production is moving towards sustainability by incorporating alternative cementitious materials, employing carbon capture techniques, utilizing recycled concrete aggregates, and developing geopolymer concrete. These innovations are leading to an increasing number of sustainable products entering the market. Planners are expected to stay informed about these developments to meet building sustainability criteria effectively, using all available technologies.

The Technological Challenge for a Holistic Approach to Sustainable Design

Achieving sustainable design in real-world projects involves balancing and integrating various of these approaches. This holistic approach presents significant challenges, as it requires extensive carbon footprint and sustainability assessments of different design variations. These assessments must be carefully balanced against other project parameters such as cost, construction complexity, and material availability during the design phase. Additionally, communication among all project stakeholders must be exceptionally efficient to ensure that sustainability investigations remain within the project's timeline and budget. Such efficiency can only be achieved through the effective use of technology, in order to automate and simplify these processes as much as possible.

3 The Digital Transformation in Construction

To illustrate the role of AI in a specific process, picture a chocolate factory equipped with an automated production line. Despite the automation, human intervention is essential at critical stages for quality control and decision-making. Similarly, AI functions like these human workers, integrated only at crucial points within a largely automated system when human-like decision making is necessary. To leverage AI for enhancing the sustainability of building structures, it is essential to grasp the current digitalization and automation. How are buildings designed, and how do engineers ensure their structural integrity in advance? How is such a design effectively translated into detailed documentation that guides construction? To what extent are construction sites automated, and how are new building materials developed? Lastly, how is communication managed both within and between these various processes?

Balancing Sketches, CAD and Algorithms in Building Design

Conceptual design of building structures can become a challenging task. For instance, the Golden Lion award for the best individual participant at the 2016 Venice Biennale—event renowned for the showcase of innovative and experimental architectural concepts and structures—was awarded to the Paraguayan "Gabinete de Arquitectura" office represented by their structurally complex masonry vault pavilion.[1] Visitors, often marveling aloud beneath the structure, questioned how such a construction could remain standing. So, how do such structures come to be?

[1] Our work as structural engineers at Fhecor Ingenieros involved the revision of the initial design of the architect Solano Benítez, in order to check the structure and make its construction feasible.

DOI: 10.1201/9781003506324-3

The initial stage of conceptual building design typically begins with a handmade sketch—usually to scale—that outlines how the building or structure will meet the intended requirements or purpose. This practice of sketching has remained unchanged for over five thousand years, regardless of whether the sketches were originally carved in stone, or are nowadays drawn with paper and pencil, or created on an iPad app. It is rather the evolution of that initial concept into a detailed design and the assurance that the building developed from that sketch will remain structurally sound that has changed over time - from the ancient empirical approaches to the complex non-linear geometric, material, and dynamic seismic calculations[2] necessary to guarantee the stability of that vault at the Venice Biennale.

For millennia, builders relied on geometric principles derived from prior constructions—and a fair amount of faith—to ensure their buildings' stability. Whether constructing Egyptian pyramids, Roman bridges, or Gothic cathedrals, they mainly replicated and refined successful past designs. While this approach was impressive for its era, as evidenced by the imposing enduring medieval and ancient structures, it inherently restricted the variety of possible designs. It was not until the Scientific Revolution that Galileo, in his 1638 work 'Two New Sciences',[3] established the foundational concepts of structural theory, detailing how to evaluate a structure's capacity based on its shape, loads, and materials. After this, structural analysis advanced rapidly with new mathematical and scientific progress. Yet, the design possibilities remained limited—constrained by available construction techniques and materials on one side, and by the limitations of manual analytical calculations suited only for idealized regular structures on the other. It was in 1930 when American engineer Hardy Cross introduced an iterative numerical analysis method that allowed to compute more complex and irregular building frames and bridges. In the subsequent years, the growing emphasis on numerical analysis resulted in the creation and eventual implementation of the

---

2 Building code regulations require a seismic analysis for all building projects located in areas prone to earthquakes considering the statistically worst possible earthquake in that area. Buildings are therefore expected to be safe for its occupants even in such events. The majority of building collapses appearing on the media after earthquake catastrophes are due to flaws in the design or construction that did not comply with the regulations.

3 Galileo Galilei's publication, "Two New Sciences," significantly advanced the understanding of structural strength and the behavior of materials, laying important groundwork for the modern theory of structures. He introduced fundamental concepts related to the strength of beams, investigating how their size and material affect their ability to withstand forces and bending moments. These contributions were made towards the end of his life, under the constraints of house arrest imposed by the Inquisition, reflecting his enduring commitment to scientific exploration even in the face of severe personal and professional adversity.

Finite Element Method (FEM) in computers starting in the 1960s. This method, still widely used today, can manage any complex structures and materials by dividing a structure into a network of interconnected small elements.[4] In fact, modern commercial structural analysis software typically includes a computer-aided design (CAD) interface where users can design the geometry, specify materials, and apply loads to a structure. This information is processed by a sophisticated (Finite Element Method) FEM engine that conducts the structural analysis to ensure compliance with stability and serviceability standards outlined in structural codes. Depending on the size and complexity of the structure, this analysis can take anywhere from a few seconds to several hours. Even so, engineers are still encouraged to carry out simplified hand calculations to verify the FEM software results, guarding against possible input errors in this increasingly complex software.

Regarding the refinement of the initial conceptual design, digitalization also plays a significant role today. As previously mentioned, the design process typically begins with architects, often working alongside engineers, who initially sketch their concepts by hand. These drawings are sometimes supplemented by scaled physical models, which help to explore the spatial relationships, forms, and lighting of the proposed structures, serving as tools for both refining design and presenting to clients. In the 1980s, advancements in computer software led to the development of early three-dimensional (3D) modeling programs such as Archicad and AutoCAD 3D. These programs enabled architects to transform their hand-drawn designs into simple digital 3D models, complementing traditional physical models and providing a more adaptable method to explore and refine their creations. In the 1990s, architecture began to embrace free forms - shapes that are flowing, irregular, and non-linear, standing in contrast to traditional geometric forms. Architect Frank Gehry was a pioneer in this movement, with works like the "Golden Fish" sculpture in Barcelona in 1992 and the Guggenheim Museum in Bilbao few years later showcasing these patterns. Due to the lack of suitable software for managing free forms in architecture, Gehry's studio opted to use CATIA, a program originally developed for aircraft design. Similarly, Zaha Hadid's studio, known for its organic, free-form architecture, adopted Autodesk Maya, a software commonly used in

[4] Ray W. Clough introduced FEM in his presentation, "The Finite Element Method in Plane Stress Analysis," at the 2nd American Society of Civil Engineers (ASCE) Conference on Electronic Computation in 1960, Pittsburgh, Pennsylvania (Clough 1960). Originally developed for airplane wing structural analysis, FEM has expanded its utility across various engineering disciplines, including structural engineering. FEM now supports the non-linear and dynamic analyses, enhancing its functionality in both research and practical applications.

the film animation industry. Simultaneously, the emerging Rhinoceros (Rhino) software quickly recognized the potential of 3D modeling tools in architecture, investing in this area and becoming a key tool for complex design in the industry (Canizares 2024). In subsequent years, computer algorithms began to address the challenges of complex geometries in architecture. For example, the geometry and triangulation of the Great Court of the British Museum, completed in 2000, were defined using a C++ script (Williams 2001). Likewise, the paneling of the unbuilt Mariinsky II Theater's free-form roof in Saint Petersburg, envisioned in 2003, was executed using a Visual Basic script (DETAIL 2013). The use of computer algorithms for defining complex geometries became significantly more accessible with the release of the Grasshopper plug-in for Rhinoceros in 2007, initially called Explicit History. This plug-in provides a visual programming environment where users can quickly develop algorithms without coding knowledge, by connecting components or pre-compiled simple algorithms to automate tasks in Rhino as shown in Figure 1.

Grasshopper enables users to create and compile custom plug-ins that enhance the standard components, expanding their functionality. Soon, third-party developers recognized the potential of merging this flexible visual programming tool with object-oriented programming (OOP)[5] to create plug-ins that go beyond simple geometry generation. One of the first plug-ins using this approach was karamba3d (Preisinger 2013), designed for algorithmic structural analysis. It serves as a FEM engine

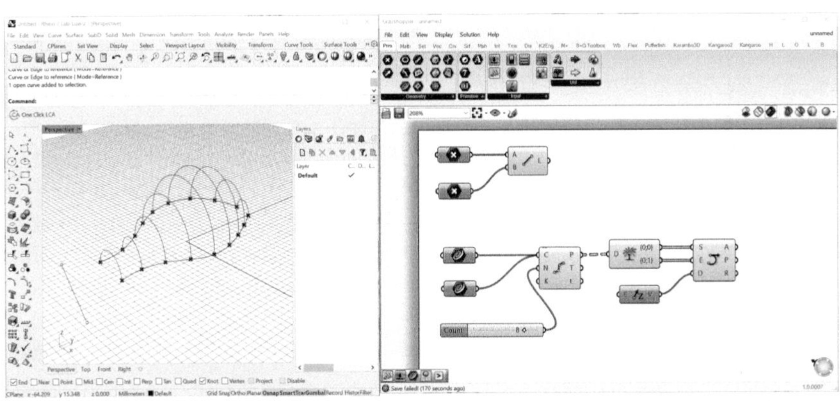

Figure 1. Simple Grasshopper script (right) to draw geometry in Rhino (left) using algorithms defined through visual programming.

[5] OOP is a programming paradigm that uses "objects"—data structures consisting of data fields and methods together with their interactions—to design applications and computer programs.

within Grasshopper, facilitating the automated analysis of structurally complex systems defined algorithmically—or parametrically—in Grasshopper, allowing for rapid iteration and optimization. Similarly, various other plug-ins have emerged to incorporate simulation tools for aspects like daylight analysis, thermal comfort, and acoustics within the Grasshopper environment. It is also feasible to develop plug-ins by leveraging application programming interfaces (APIs) from standalone commercial software (Apellániz and Vierlinger 2022), rather than embedding simulation engines directly into the Grasshopper plug-in. This approach was also used by Apellániz, Pasanen, and Gengnagel (2021) to integrate commercial LCA software into Grasshopper, automating this environmental analysis with a parametric 3D model. Currently, the official "food4rhino" platform hosts over 1000 plug-ins, many available for free, with millions of downloads. Although public interest in freeform and complex geometries has declined recently due to rising construction costs and environmental concerns about material usage, these tools, workflows, and automations are still valuable for swiftly analyzing and iterating through different design variations of more standardized buildings, thereby enhancing the final design.

Project Documentation with BIM

The main aim of any construction project is to be completed, which requires the design to be thoroughly documented. These crucial documents are the basis for the tendering process—where contractors submit proposals to carry out the construction—and they also guide the actual building work. They generally include architectural, structural, and MEP drawings, as well as a bill of quantities and detailed specifications outlining the required quality, materials, and workmanship standards for each part of the construction. Although these documents can differ from one country to another, their accuracy is vital to avoid costly mistakes. Such errors could lead to legal actions from contractors[6] claiming discrepancies that either obstruct construction or raise project expenses—sometimes justifiably.

Traditionally, architectural and engineering drawings were crafted as hand-drawn blueprints. These were produced on large drafting tables

[6] As an example, the Panama Canal expansion which started in 2016, while a monumental feat of engineering, was plagued by extensive legal disputes primarily concerning cost overruns. Grupo Unidos por el Canal (GUPC), the construction consortium, engaged in protracted arbitration with the Panama Canal Authority (ACP), arguing that unforeseen geological conditions not documented in the project specifications led to billions of dollars in extra expenses. These disputes resulted in numerous arbitration awards, with varying financial outcomes for both parties. Essentially, the litigation centered on the allocation of financial responsibility for the project's significant cost escalations.

using tools like T-squares, drafting machines, pencils, and ink. The process demanded careful attention to detail and often involved lengthy revisions. With the rise of personal computers and the advent of CAD software such as AutoCAD[7] in the early 1980s, the creation of drawings transitioned from manual to digital. AutoCAD and similar software facilitated the more precise and efficient production of technical drawings, allowing for easier modifications and reproductions. Meanwhile, the bill of quantities was typically prepared by a quantity surveyor, who manually calculated quantities from the drawings to compile a comprehensive list used for cost estimation and tendering. Additionally, specifications are written documents usually prepared by architects and engineers who often reuse entries from previous projects for consistency and efficiency. A challenge arose because all these technical documents, even when digitized to some degree, were created independently, potentially leading to coordination issues. This could lead to coordination issues, such as undetected clashes between MEP and structural drawings, calculation errors in the bill of quantities, or missing specifications for certain materials.

Since the 2000s, the project documentation process has undergone significant changes due to the widespread adoption of Building Information Modeling (BIM),[8] though its development began back in the 1970s. BIM is a collaborative approach designed to create and manage comprehensive information about a construction project throughout its lifecycle. It employs 3D models to visualize buildings and infrastructure, as shown in Figure 2, enhancing collaboration and decision-making. The model's geometry is systematically categorized into logical families and types, providing detailed information about each element, such as walls, floors, beams, materials, levels, area, volume, and weight. BIM offers three major advantages. Firstly, it enables the generation of drawings, bills of quantities, and specifications directly from a unique BIM model, ensuring consistency across all documents. Additionally, during the design phase, BIM models from various disciplines are regularly coordinated to detect

---

[7] AutoCAD, one of the most pivotal CAD programs, was introduced in 1982 by Autodesk, co-founded by John Walker. However, the invention of CAD itself is often credited to Ivan Sutherland, who developed Sketchpad in 1963, a program that laid the foundational concepts for today's CAD software.

[8] The concept of a 'building model,' similar to today's BIM, was initially introduced in a 1985 paper by Simon Ruffle, published in 1986. The phrase 'Building Information Model' was first used in a 1992 paper by G.A. van Nederveen and F. P. Tolman. It took nearly a decade for the terms 'Building Information Model' and 'Building Information Modeling' to gain widespread acceptance. This shift occurred around 2002 when Autodesk popularized the term through a white paper titled "Building Information Modeling," prompting other software vendors to highlight their contributions to the field.

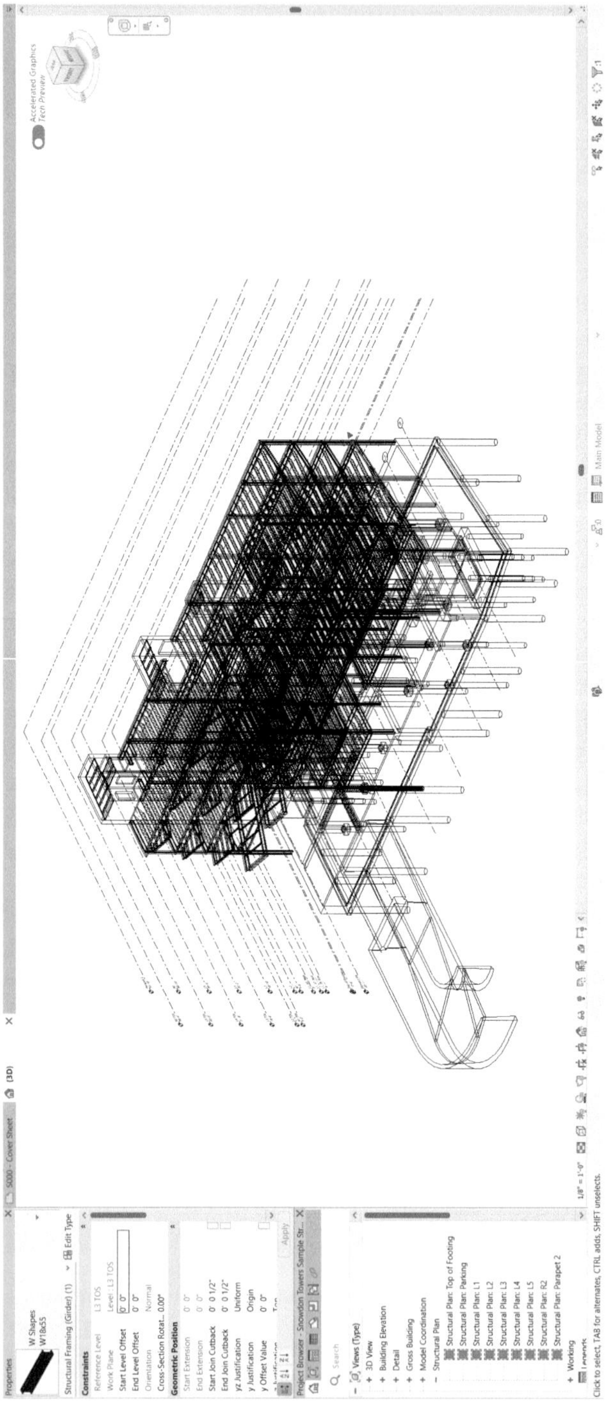

Figure 2. Structural BIM model created with the program Autodesk Revit.

clashes and compatibility issues, ensuring consistency among the different sets of drawings. Secondly, because BIM models are 3D and information-rich, they can serve as input for building simulation tools during the design phase. Thirdly, BIM's centralized information and detailed specifications enhance facility management by providing data about the building's structure, facilitating future renovations, and aiding HVAC maintenance scheduling. Furthermore, BIM models can be integrated with dynamic data from Internet of Things (IoT) sensors and other tools during the building's operational phase, creating a digital twin that monitors various aspects of the building in real time.

Implementing BIM comes with several challenges. These projects often incur higher costs due to necessary investments in software, employee training, and the expenses related to collaboration among all planning disciplines. Thus, there must be a demand from clients who are willing to pay these additional costs to benefit from improved project document coordination and enhanced facility management. Many administrations are already mandating BIM use in public projects for these reasons. Regarding BIM collaboration, there are two main approaches, each with its own set of challenges. Closed BIM, or proprietary BIM, requires all stakeholders to use the same software, which can lead to software dependency and compatibility issues with external parties. This approach may limit flexibility and increase costs due to licensing fees, although it enhances consistency within the project team. On the other hand, Open BIM encourages interoperability by using open standards like universal Industry Foundation Classes (IFC) files, allowing stakeholders to employ different software tools. However, it demands strict adherence to modeling standards to maintain data accuracy and utility, which can be difficult to enforce across diverse teams. Last but not least, BIM is typically applied in the later design stages due to the complexity of BIM modeling compared to traditional 3D modeling programs. This makes it challenging to use BIM in early comparative design analysis, resulting in the possibility that BIM models may not be ready for building simulations needed for informed design decisions during these studies.

Advances in Materials Science and Automation in Construction Sites

Throughout history, the discovery of new building materials has sparked major transformations in construction. About 5000 years ago, the use of stone and masonry enabled the construction of lasting monumental structures and temples, establishing stone as one of the most durable and sustainable building materials. During the 18th century, as the Industrial

Revolution unfolded, iron and later steel began to be incorporated into construction. These materials, known for their enhanced strength and durability, allowed for the creation of longer bridges and taller buildings. The late 19th and early 20th centuries saw the emergence of reinforced concrete, which combined steel rebars for tensile strength with concrete for compression, enabling innovative building and bridge designs. The development of high-strength concrete throughout the 20th century revolutionized skyscraper construction. More recently, advances in wood engineering have introduced high-strength engineered wood products like CLT and LVL, making it feasible to construct multistory buildings and modest bridges using timber. Today, material science is focused on developing building materials with superior mechanical properties and a reduced carbon footprint, potentially leading to another revolution in construction. To this end, research institutions are actively exploring, synthesizing, and analyzing new material structures, often utilizing cutting-edge technologies such as nanotechnology and computational modeling.

In addition to researching new materials, there is also an emphasis on innovating the manufacturing process of construction materials—key focuses include reducing energy consumption and incorporating recycled materials. Once these building materials are produced, they can either be sent directly to the construction site for on-site construction—such as pouring fresh concrete—or used in the production of prefabricated elements, such as CLT panels for multi-story buildings and precast concrete beams and columns for industrial structures. Significant innovation is taking place in prefabricated products, primarily to shorten construction time and, consequently, lower labor costs. However, new prefabricated products usually need certification before they can be utilized, and the process of testing and proving their viability to certification agencies can be arduous. A recent trend gaining traction in prefabrication is modular construction, mainly involving timber and steel. This involves assembling entire building modules, which can include one or more rooms complete with structure, finishes, and MEP systems, before they are shipped to the site for assembly. Conversely, construction sites tend to be hesitant to adopt innovations. The most notable progress in recent years has been in enhancing safety management and hazard prevention and control, with a significant disparity between developed and developing or undeveloped countries. Some recent advancements are more the exception than the norm on construction sites. For instance, there has been substantial investment in developing 3D printing technologies for construction (see Figure 3), and some companies have created robots to perform specific tasks like drywall finishing. Additionally, advancements in augmented

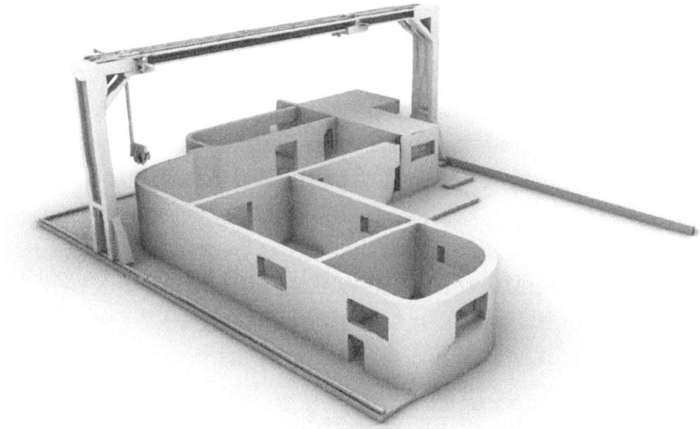

Figure 3. 3D concrete printing in construction.

reality (AR)[9] offer numerous possibilities for ensuring accurate placement and construction of all elements on site. Some projects have even been completed without printed drawings, relying solely on a BIM model and AR for construction coordination.[10] Finally, there is a growing push, particularly in Scandinavian countries, for electric construction machinery, especially on municipal building sites, to reduce emissions and noise pollution (Henley 2025).

Project Communication in the Post-Pandemic Era

Construction projects are inherently complex tasks involving multiple stakeholders, extensive coordination, and dynamic interactions across various disciplines. Planners and designers collaborate closely to ensure design consistency, technical feasibility, and interdisciplinary coordination. As a part of this collaboration, they work with the building owner or developer to clarify project goals, agree on design proposals, manage the budget, and more. Simultaneously, the building owner must interact with regulatory authorities to secure necessary permits, ensuring compliance with local laws and regulations. Furthermore, planners and designers engage with contractors to discuss construction feasibility and

[9] AR is a technology that overlays digital information—such as images, models, or data—onto real-world environments in real-time, typically viewed through devices like smartphones, tablets, or AR glasses.

[10] The construction of the new Stavanger University Hospital in Norway adopted a paperless approach, employing model-based construction processes to enhance digitalization and innovation. However, challenges such as high costs of creating high-quality BIM models and interoperability issues between software programs were encountered (Disney et al. 2023).

interpret design details. These interactions are just a few examples of the extensive communication required in construction projects. Projects with sustainability objectives demand even more intensive communication. In these cases, planners must delve deeper into discussions about design options and present them to the client, ensuring that all aspects of sustainability are considered. When an external consultant is involved to achieve a green certification, such as LEED or BREEAM, their participation becomes crucial. They must join discussions between planners and the client to address the specific requirements of the desired certification and ensure that the project aligns with sustainability goals from start to finish. During the construction phase, it is imperative to verify that the materials used and the construction practices adhere to the agreed-upon carbon footprint and sustainability criteria established during the design phase. Finally, all relevant data must be meticulously compiled and submitted to the certification agency for approval, completing the cycle of rigorous oversight and management required for sustainable building projects.

Traditionally, communication in construction projects depended heavily on in-person meetings, which included frequent visits to sites and direct discussions. However, these meetings often proved impractical, particularly due to the often remote locations of construction sites, leading to logistical issues and inefficiencies. The Coronavirus disease 2019 (COVID-19) pandemic significantly altered this scenario when employees had to work from home suddenly. It highlighted the necessity for collaborative tools. Companies had to quickly adopt or speed up the use of digital communication platforms like Microsoft Teams, SharePoint, and other collaborative tools. These were used not only for internal communication among colleagues and accessing files through cloud solutions, which generally received a positive response (Hiti et al. 2021), but also for external communication with other project stakeholders, allowing meetings to take place online instead of in person. This shift allowed stakeholders to maintain consistent and effective communication remotely, overcoming geographical limitations and greatly enhancing operational flexibility and efficiency.

The successful adoption of online collaboration has sparked a more favorable attitude towards using various cloud-based tools in construction projects. This technological shift began with platforms like the Autodesk Construction Cloud, where designers uploaded BIM models to a central file-sharing system for coordination via internet browsers, thus eliminating the need for emailing model files. Following this advancement, AR gained traction, offering new applications that enabled construction teams to use AR for inspections and detailed site assessments. This approach significantly reduced the necessity for on-site visits and allowed stakeholders to view project progress remotely, enhancing efficiency and engagement. Furthermore, some architectural firms, including BIG, Zaha Hadid Architects, and Foster + Partners, are exploring the metaverse[11] for immersive presentations, collaborative design, and real-time client feedback. By integrating these innovative technologies, these firms are not only transforming professional practices but also redefining how building projects are visualized, developed, and experienced.

[11] The metaverse refers to a collective, immersive virtual space combining AR, virtual reality (VR), and the internet, enabling users to interact, collaborate, and experience digital environments similarly to physical spaces.

Machine Learning and Big Data

4

While the use of large data sets for training machine learning (ML) models has been a standard practice in many industries for years, its application in the construction sector has been relatively limited. This chapter opens with a brief introduction to ML algorithms, followed by an interview with Alexander Stirken, a structural engineer turned software developer with whom I have had the opportunity to work. Alexander offers insights into both the challenges and the untapped potential of software development and ML in transforming the construction industry and reducing its carbon emissions. These initial sections set the stage for a review and discussion of how current research and the application of ML in specific areas of the building industry are paving the way for more sustainable construction. The concluding section discusses a project from my doctoral thesis (Apellániz 2024) that aimed to bring these workflows closer to everyday professional practice.

Introduction to Machine Learning

ML refers to computer algorithms that enable computers to perform certain tasks without explicit instructions on how to do so. Instead, these algorithms are exposed to data from which they can identify patterns and make predictions or decisions based on those patterns. ML is considered a subset of AI. While AI is a broader concept that encompasses any technique that enables computers to mimic human behavior, ML specifically involves computers learning from data.

ML algorithms are generally categorized into three types: supervised, unsupervised, and reinforcement learning (see Figure 1). Supervised learning involves training algorithms on a labeled dataset, where the correct outputs are known, allowing the algorithm to learn over time to

DOI: 10.1201/9781003506324-4

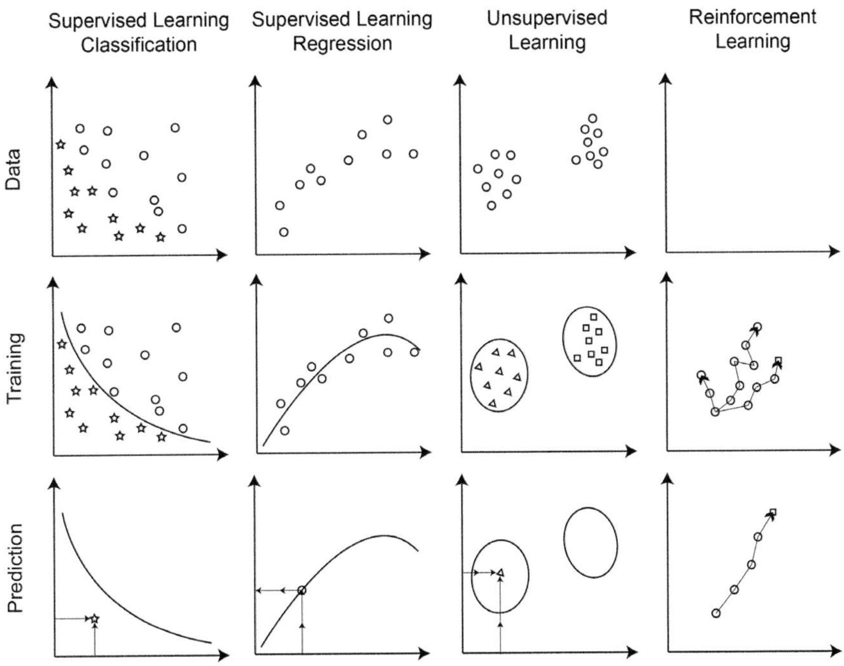

Figure 1. ML algorithms.

predict outcomes based on the provided features. It includes tasks like regression (predicting continuous values) and classification (categorizing data into predefined groups). Unsupervised learning, on the other hand, uses data without labeled responses and aims to identify patterns or structures based on the data features; common applications include clustering and association. Reinforcement learning is significantly different from the previous algorithms. It focuses on training an intelligent agent—a system such as a robot or software acting autonomously—to take actions to achieve a specific goal. Unlike other ML methods where a model learns from a pre-existing dataset, a reinforcement learning agent learns by interacting with an environment where no predefined data is available. The algorithm learns from the outcomes of its actions, receiving rewards for successful actions and penalties for unsuccessful ones, thus gradually improving its decisions over time. Due to the particular characteristics of reinforcement learning and its synergy with generative AI methods, which have not been covered yet, it will be specifically addressed in a separate chapter.

ML algorithms can also be categorized into statistical methods and neural networks, according to their foundational principles and uses. Statistical methods consist of explicit mathematical models with relatively few parameters, often based on strong assumptions about the data (such as linearity, independence, or normality). Neural networks, on the other hand, consist of multiple layers of interconnected nodes with potentially thousands or millions of parameters, which allows them to approximate highly complex and nonlinear functions. While statistical methods are still widely used in many practical applications, particularly when the dataset is small or medium-sized, the relationships in the data are relatively simple or the interpretability of the predictions is important; today, neural networks are more prominent, especially in fields where very complex patterns need to be learned from large amounts of data—such as image recognition, natural language processing, or autonomous driving. ML algorithms employing multi-layer neural networks are also called deep learning algorithms.[1]

To explain neural networks at a basic level, let's start with a perceptron (see Figure 2), which is the simplest neural network possible: a computational model of a single neuron. Its behavior can be explained through simple arithmetic. The numerical inputs x_1, x_2, ..., x_n are each multiplied by corresponding weights w_1, w_2, ..., w_n. These weighted inputs are then summed together along with a bias term b, forming the weighted sum $y = \sum w_i x_i + b$. This sum y is then passed through an activation function σ, which transforms y into the output $\hat{y}$. If the parameters of the model—the weights and the biases—are properly tuned, a perceptron could be applied to perform simple binary classification tasks. For instance, the input could be several financial indicators of an investment project, and the output of the model would indicate if the project is worth investing into if ($\hat{y} > 0$) or not ($\hat{y} = 0$). The tuning of the model parameters is called

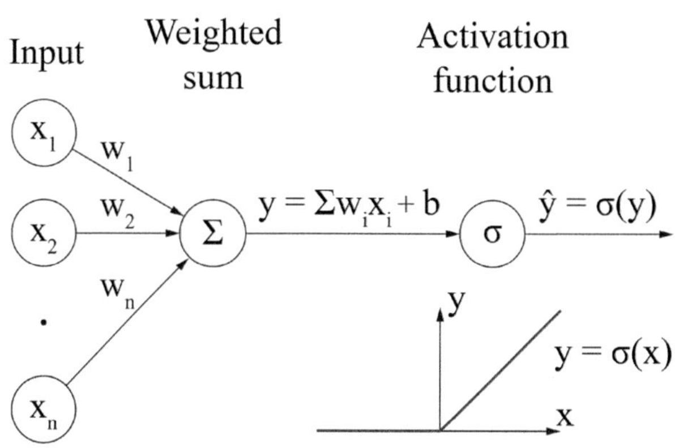

Figure 2. Perceptron.

[1] Strictly speaking, deep learning refers to the use of multilayered neural networks, however the vast majority of neural networks used in practical applications fall into this category.

training. In this example, a database of previous investment projects and their resulting success would be necessary. The model parameters are then adjusted so the result $\hat{y}$ after feeding the indicators of the recorded projects into the model matches as accurately as possible the recorded success they had. After training, the perceptron uses these tuned parameters to predict the expected success for new, similar investing projects, effectively generalizing from the examples it was trained on to new situations.

The perceptron concept can be expanded as shown in Figure 3 by layering multiple neurons in interconnected layers—input, hidden, and output—enabling the neural network to perform more complex tasks due to a higher number of parameters. These parameters allow the model to capture a broader variety of features and relationships within the data. For example, in a facial recognition task, the input image is transformed into numeric data that can be fed into the model based on the color of the different pixels, the first layers can then learn to recognize different facial features in the input layer, aggregate these features in later hidden layers, and finally classify the identity in the output layer. However, as the depth and complexity of the network increase, it becomes increasingly difficult to interpret and track how the neural network processes the input and arrives at its predictions, hence their reputation as "black box" algorithms.

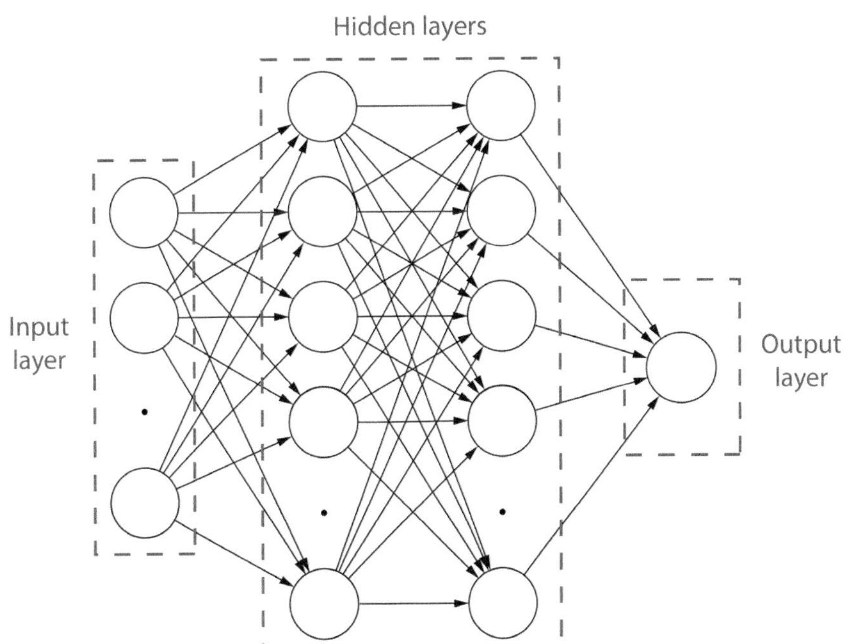

Figure 3. Dense neural network.

Typically, the more complex the task to be solved is, the more parameters the model must have. For hand-written numbers recognition, a neural model with around 100000 parameters may suffice, while some LLMs—neural networks trained to understand and reproduce natural language that will be further explored in later chapters—may consist of up to trillions of parameters. Besides the number of parameters, the so-called architecture of the neural network is fundamental to the problem to be solved. Neural network architecture defines the structure of layers and neurons in a model, organizing how data flows and is processed to learn from inputs and make predictions. While a dense neural network like the one of Figure 3 is characterized by every neuron in one layer being connected to every neuron in the next layer, with each connection processing data in the same simple approach as the perceptron, an LLM like ChatGPT utilizes a far more complex transformer architecture, which relies on self-attention mechanisms to process sequences of data in parallel inside each layer, as presented in the acclaimed research paper "Attention is All you Need" (Vaswani et al. 2017). Nevertheless, this chapter concentrates on ML approaches, which involve gathering targeted big data and training algorithms to address those concrete challenges of the building industry. General-purpose pre-trained models, such as LLMs, are discussed in following chapters.

A Discussion with Alexander Stirken on Software Development in the Building Industry and the Challenges and Potentials of Machine Learning

DA: We both studied structural engineering but ended up working in software development within the construction sector. Software developers in this field often have academic backgrounds in civil engineering or architecture, rather than in traditional software development. Could you share a summary of your career journey and how you shifted from structural engineering to software development?

AS: I earned a Bachelor of Science (BSc) in civil engineering and worked at a bridge design company focusing on concept and detail design. Later, I pursued an Master of Science (MSc) in structural engineering at Hannover while simultaneously working for Deutsche Bahn.[2] During my studies, I took a BIM course, but aside from Matlab, I didn't have much exposure to programming. My first real encounter with programming

[2] Deutsche Bahn is Germany's national railway company, responsible for operating the country's passenger and freight rail services, as well as managing much of its rail infrastructure.

came during my master's thesis, where I self-taught the necessary skills to create an interface between the BIM program Revit and the FEM program Sofistik, automating the creation of calculation models from BIM models—a relatively new concept at the time. Following my advisor's recommendation, I remained at the university after graduation to pursue a PhD, focusing on programming. The university department I joined was centered around BIM implementation in professional practice, a cutting-edge topic then. What I found fascinating about programming was the instant feedback received when clicking the "debug" button, unlike the lengthy process of completing railway bridge projects at Deutsche Bahn. Basically, I taught myself programming through online resources and worked on side projects. However, due to personal reasons, I returned to my hometown of Bielefeld before finishing my PhD and secured a position as a software developer at Goldbeck,[3] where I felt I could grow and learn from others.

DA: At such a big company you probably worked with colleagues that were both either software developers with that academic background or self-taught programmers with a background as architects or engineers like yourself. What would you say were your strengths and weaknesses having your background in structural engineering?

AS: So, my official position when I got hired at Goldbeck was "BIM software engineer". Our department was divided into two parts so to speak: People that worked on processes and workflows related to architecture and engineering (who were engineers and architects that wanted to work in digital workflows) and people that did pure programming who were mostly computer scientists. I noticed that it was really important to speak both languages. For instance, when computer scientists were confronted with "rebar data" or "punching reinforcement checks", they were overwhelmed; and on the other hand, if a computer scientist was talking about pure Information technology (IT) terminology like APIs, software architecture, web frameworks, etc. it was difficult to discuss these topics with people working purely on BIM workflows. Since I spent two whole years focusing solely on this IT domain (which may be difficult for many civil engineers when they come across web sockets, frameworks, etc., for

[3] Goldbeck is one of Europe's biggest construction companies, with over 12,000 employees in more than 100 locations, known for delivering commercial and public buildings using a highly industrialized and digitalized approach known as the "Goldbeck system". The business model is that they offer the whole package vertically integrated from the first selling talks with the clients, then design - architecture, structure, MEP—to the preassembly of the building elements (mostly structure) in the factory to assembly in the construction site legowise. All this is a relatively rapid process subjected to preestablished prices.

the first time), I could notice how valuable it was to be versed in both worlds.

DA: Goldbeck is one of the market leaders in the design and construction of above-ground parking buildings. You played a significant part in developing the parking configurator application at Goldbeck, which is now used to configure its parking buildings. Can you describe your role in this process?

AS: Goldbeck specializes in systematic construction, with roughly ten percent of its market dedicated to parking buildings—a typology that is highly standardized. At Goldbeck, there is one common type of parking structure that can be adapted to nearly every situation, characterized by fixed specifications such as standard parking widths, level heights, defined column and beam spacings, etc. This level of standardization sparked the idea within the company to develop a configurator capable of automatically generating parking buildings based on specific project requirements. While the company had already developed some configurators, like one for trusses,[4] this parking building configurator marked the first time they tackled an entire building. Its viability seemed promising, given the inherent systematic nature of parking structures compared to other building types like residential or commercial projects. Architects and engineers would begin by defining the main parameters of the building—such as the number of parking spaces, ramps, staircases, entries, and exits—using numeric and semantic inputs constrained to the configurator's possibilities. This controlled input ensured that users could not enter parameters that would result in a design incompatible with the Goldbeck system for preassembling and constructing parking structures. From there, the configurator would handle the detailed design aspects, like the position of the parking spaces, the elevation of the levels in the building, the type of levels (whether split levels or full stories), etc. The output was a BIM structural model, which architects could later use as a base to manually adjust the facade and other nonstructural elements according to the parking building's surroundings and aesthetic preferences. The configurator was developed as a web tool and deployed on a powerful external server to handle the computationally intensive model generation process. From the BIM model, workshop drawings were automatically generated for factory production. An important requirement of the configurator was its accuracy: Every structural element needed to have the correct size and position to avoid any issues during assembly and construction. The project was initiated with a minimum

[4] A truss is an assembly of beams, connected creating a triangular pattern, that creates a rigid structure to support roofs, bridges, etc.

viable product (MVP),[5] and early users were somewhat skeptical because certain configurations achievable with the Goldbeck system could not be generated due to limitations in the algorithm. In essence, users were initially hesitant to trust a "black box" algorithm—even without any AI involved—to produce an accurate model.

DA: I presume the parking configurator's algorithm was designed to create the building following specific coded rules. Considering Goldbeck's size, they likely have hundreds of BIM models from past parking structures. Did you ever think about using AI by training a neural network on those designs to produce the parking models, or was the decision to manually write all the rules for the generative algorithm the most obvious choice from the start?

AS: The parking configurator was exactly as you describe: a fixed algorithm capable of generating all possible variations of parking structures for a given input while adhering to the Goldbeck system. Initially, during its development, there was consideration of training an AI using all the BIM models we already had. Training data was prepared by associating each model (the outcome) with the hypothetical input parameters that would produce it in the configurator. We first attempted statistical models like random forest and nearest neighbor, followed by an AI model using an artificial neural network. Unfortunately, the training process was unsuccessful. The standard deviation for the statistical models was excessively high, and the neural network's training loss failed to converge. This likely indicated that the training data, comprised of 500 models solely of parking buildings, was insufficient. Although generating additional synthetic training data[6] could have enhanced the training process, research into the computational resources required on the Microsoft Azure cloud platform revealed that the cost was disproportionately high compared to potential benefits, even if the training succeeded. Consequently, we decided to proceed with coding the entire algorithm manually.

DA: I genuinely believe it was the right choice. Often, I get inquiries from people eager to use ML simply because they have some training data available. ML is most beneficial when dealing with complex patterns

[5] An MVP refers to the simplest version of a product that includes only the essential features needed to solve a specific problem or deliver value to users. The purpose of an MVP is to validate an idea as early as possible by releasing it to real users, gathering feedback, and using that insight to guide further development.

[6] Synthetic training data refers to artificially created data that simulates real-world data for training ML models. This approach is often used when actual data is scarce, sensitive, or expensive to obtain.

in data that aren't easily captured through traditional programming methods. It is particularly effective in areas like image recognition, natural language processing, and predictive analytics, where insights derived from data are essential. Conversely, crafting an entire algorithm from scratch is better suited for problems that are well-defined and rule-based, where outcomes and procedures can be clearly articulated. Nevertheless, collecting extensive data to train neural networks for construction projects is challenging due to the distinct and varied characteristics of each project. Factors like irregular shapes, different design requirements, construction materials, and local regulations make it hard to gather a large, uniform dataset. On the other hand, industries such as e-commerce and social media are ideal for collecting big data, thanks to the constant stream of user interactions. These industries have already effectively utilized ML for tasks such as personalized recommendations, targeted advertising, and sentiment analysis, where massive amounts of behavioral data drive continuous optimization and improvement.

AS: To begin with, as a society, we should focus on collecting more data from completed building designs and using this information early in the project design phases to make well-informed decisions. Without this, productivity remains stagnant and lags behind other industries. How can AI play a role alongside this data? One idea is through design copilots: users can set parameter limits to establish a range of potential solutions and input the project's site geometry. The algorithm then proposes designs based on past data. Another strategy involves addressing what the parking configurator did not achieve, which is predicting aspects like steel quantity and carbon footprint based on previous designs for a specific building shape before running analyses or generating the building with the configurator. This allows for quick adjustments of input parameters to enhance the design.

DA: Absolutely. Those two approaches are certainly fascinating, and they will be discussed in greater detail in later sections of the book.

Data Driven Carbon Footprint Estimation

Predicting carbon footprints can help pinpoint critical parameters affecting emissions early in the design phase, enabling architects, engineers, and decision-makers to make sustainable choices proactively rather than reactively. Since the primary aim of ML is prediction, it is logical to apply it for this purpose. However, the fundamental requirement is the availability of data on building carbon footprints to train the ML model. Figure 4 intends to depict such a database of buildings. It highlights the diversity of buildings and the challenges in defining features to represent

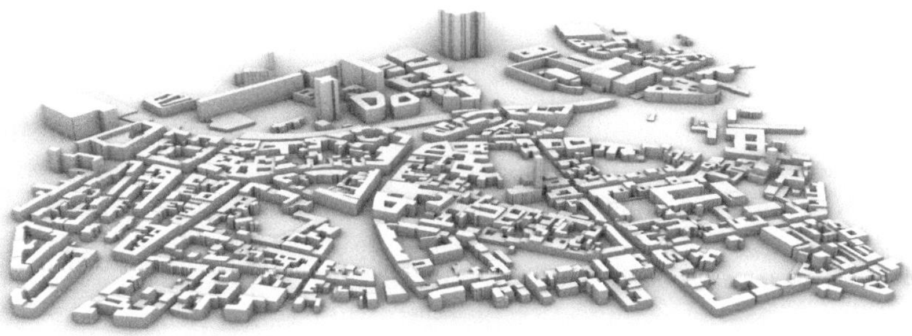

Figure 4. The built environment as database for ML models.

them. Once this data is collected and pre-processed as needed, various ML algorithms should be compared and assessed—similar to what was mentioned in the previous interview—to determine the best predictive option. Unfortunately, commercial applications that use carbon footprint prediction typically do not reveal their methods, so it is beneficial to start by examining scientific publications where this approach is detailed.

Fenton et al. (2024) present a clear and comprehensive method for predicting early-stage greenhouse gas emissions from buildings. They utilized the Embodied Carbon of European Buildings Database (EUCBD) (Röck and Sørensen 2022), which includes around 625 building cases from across Europe. This database features data fields such as building type, gross floor area, construction year, structural materials, and life-cycle embodied carbon emissions (standardized per square meter over a 50-year lifespan), which were compiled through cooperation with national databases, published case studies, industry inputs, and standardized data collection forms. The study begins by selecting two case studies from the EUCB-D database: French residential projects and combined French-Belgian projects. This selection maintains moderate heterogeneity to enhance training, while still reflecting real-world scenarios since they are actual projects. The data from these records used for training reflected typical limited early design stage information, when ML predictions would be useful: gross floor area, building use (residential, commercial, etc.), structural system, number of users, and so on. The embodied carbon values of the buildings were extracted as target values or labels for the training process.

Prior to training, it is usually necessary to format and clean the data. In this instance, quantitative data such as gross floor area was scaled between –1 and 1. This technique helps ensure that no single variable (like floor area) dominates the loss function during training. Categorical data, such as building use, must be converted into numerical form, as discussed earlier in the chapter. Here, the one-hot-encoding method was utilized,[7] which generally results in effective training outcomes. When using data from different sources, it is crucial to ensure consistency across the datasets (e.g., building area measured using the same method, carbon footprints considering the same life-cycle stages, etc.).

After preparing the data, nine different statistical regression algorithms were trained. Some buildings were excluded from the training to serve as validation data, which helps verify if the trained algorithms can accurately predict the carbon footprint of those buildings and determine which performs best. Prediction accuracy, defined as the percentage of test-set predictions that fall within ±15% of the true embodied-carbon value, was used as the evaluation metric. Depending on the algorithm and selected building features, this accuracy reached up to 85% on the validation data. The study also highlighted the significant impact of structural typology and material type features.

An advantage of using interpretable statistical methods with predefined formulas is that results can be easily explained to designers and policymakers. However, numerous studies, such as the one by Zhang et al. (2024), focus on using artificial neural networks for predicting the carbon footprint of buildings. That particular study employed 850 residential building samples. While deep learning algorithms require larger training datasets, they can achieve higher prediction accuracy when handling complex data features compared to statistical models.

Beyond the scientific community, predicting carbon footprints has become an increasingly common feature in commercial LCA software for buildings, such as One Click LCA, Tally, Athena Impact Estimator, or the Embodied Carbon in Construction Calculator (EC3). These commercial LCA tools typically reference a wide range of EPDs, enabling users to conduct a verified LCA of their buildings. This verified LCA can then be submitted to a certification agency as part of obtaining a green building certification or to conduct efficient and reliable early-stage LCAs. For example, One Click LCA encourages users to submit their building LCAs to enhance the database, which in turn improves their carbon footprint prediction and carbon benchmark capabilities. Their "embodied carbon benchmarks for European buildings" have been compiled from 15,000 voluntarily submitted building LCA projects.

Predicting carbon footprints spares users from having to detail all the materials and processes of their buildings and then match them to the appropriate EPDs for the LCA. This feature is typically employed in the

[7] With the One-Hot Encoding method, each category is represented by an array where all elements are zero except for the position corresponding to the category, which is set to one. This process simplifies the input for ML algorithms that expect numerical values without imposing an arbitrary ordinal relationship.

very early stages when no architectural sketches are available and only a comparison of the carbon footprints of different building concepts is needed. Thus, a rough estimate is sufficient, and there is no need for an accurate LCA. Another scenario where predicting a building's carbon footprint is useful is when establishing a baseline design. This baseline serves purely as a reference point against which the carbon footprint of the actual design can be optimized. It is for instance especially useful for comparing a renovation project to the baseline design of a similar new construction, which would otherwise require developing an entirely separate building design to carry out the LCA. Certification agencies often request this baseline design process when submitting the building LCA.

Interestingly, there is a growing trend of applications beyond traditional LCA software incorporating carbon footprint prediction for buildings as an added feature. One of them is Autodesk Forma, which is a cloud-based tool designed for early feasibility assessments. It includes data-driven carbon footprint prediction of buildings alongside other analyses, utilizing Autodesk's internal building databases. For these predictions, Forma employs the independent c.scale API, which uses proprietary ML models and internal databases to estimate the embodied carbon of buildings (Carbon Leadership Forum 2025). An API consists of a set of rules and protocols that allow different software systems to interact., which means that the ML-based c.scale API was developed to be integrated into various applications, potentially broadening its usefulness for any program seeking to improve building design with sustainability insights driven by data.

Surrogate Models for Fast Estimations and Energy Savings in Simulation

Similarly to the LCA, other analyses can benefit from an AI approach to provide rough but sufficiently accurate estimates in the early stages. The challenge with the LCA involves accurately modeling all materials and building processes to calculate the carbon footprint. Meanwhile, other analyses face the challenge of requiring significant computational power for rapid result availability. Surrogate models offer a solution to this issue. These are computational techniques that approximate complex simulations using ML models that predict the outcomes of those simulations. These models are generally trained with historical data from detailed simulations or real-world experiments, allowing them to deliver quick and efficient approximations without the extensive computational costs of full-scale simulations. This not only accelerates preliminary analyses, potentially leading to more sustainable designs, but also results in significant energy savings by avoiding full simulations. For instance, one of the analysis tools of the cloud-based software Autodesk Forma is wind analysis, for which it uses surrogate models as illustrated in Figure 5. It offers users the option to receive a quick AI estimation of the results, which is produced by the surrogate model, or to run a detailed analysis that might take several hours, which is an actual simulation. Since the AI prediction is provided in this case as a cloud-hosted Software as a Service (SaaS),[8] a single optimized server performs the lightweight surrogate-model inference for many users, eliminating the need for countless full simulations on individual personal computers (PCs) and amplifying the overall energy savings.

It's important to note that while AI-based tools can provide users with complex analysis results, users are still responsible for interpreting and applying these results in building design. When dealing with surrogate models, users need to validate results on two levels. First, they must determine if the simulation method being used as a surrogate is reliable for a specific application. For example, a certain globally recognized structural engineer only trusts wind tunnel tests for determining wind forces on skyscrapers completely dismisses the use of computational fluid dynamics (CFD) for this task—the same method that Forma uses for wind

Figure 5. Autodesk Forma software's AI-based wind analysis.

[8] SaaS is a software-delivery model in which an application runs on cloud servers operated by the vendor and is accessed by users over the internet, typically through a web browser. Instead of installing and maintaining programs locally, subscribers simply log in to use the latest version, with all computation, data storage, security updates, and scaling handled centrally.

analysis to estimate user comfort. The key of course is to know the goal and the limits of each simulation approach. The second validation level concerns the AI model's accuracy in predicting simulation outcomes. To facilitate this, it would be beneficial if the model's training accuracy and the data used for training were made publicly available by software distributors, allowing users to judge whether a particular AI surrogate model is suitable for their project.

Surrogate models also hold significant promise in structural analysis, particularly for simulations that require substantial computational resources, such as dynamic earthquake analysis (Gudipati and Cha 2021) or soil-structure interaction analysis (Ruiz López et al. 2024). However, despite their potential, this approach has not yet been incorporated into any major commercial FEM software. One likely reason for this is the challenge of relying on AI predictions to produce structural designs, given the safety concerns involved, which differ from using surrogate models in comfort wind analysis where inaccuracies are less critical. Nevertheless, surrogate models could be effectively employed for initial comparisons of various structural designs, with the final design being verified using an actual FEM simulation.

Finally, Physics-Informed Neural Networks (PINNs) embed the governing physics in the learning process by penalizing deviations from the underlying partial differential equations, creating surrogate models that generalize well even with limited training data. As demonstrated by Zhou and Ye (2024), PINNs can be trained to efficiently simulate (structurally efficient) form-finding even on modest computational hardware.

Application to Structural Health Monitoring to Extend the Life-Cycle of Structures

This section deals with enhancing a building's sustainability by extending its theoretical service life, thus postponing the need for demolition and reconstruction. All building structures are conceived with a design service life of about 50 years for ordinary buildings. Over time, unavoidable wear of materials, accumulated strain, and changing environmental and load conditions diminish their original safety margins.

The first step toward extending a building's service life is ensuring the proper design and construction of the original building. For example, improperly placed rebars can lead to insufficient concrete cover, making steel vulnerable to carbonation or chlorides and causing premature corrosion. Similarly, timber details that trap water like flat ledges, unventilated cavities or end-grain exposure can hold a high level of moisture and start fungal decay. However, if a building can still prove its reliability at the end of its intended lifespan, there is no reason not to extend its use, although a careful assessment is mandatory.

In ordinary buildings, the life-extension assessment typically follows the ISO 13822 path: review of drawings and records, a thorough visual survey (locally exposing critical elements where needed), non-destructive testing, and, where material properties or potential defects remain uncertain, targeted sampling for laboratory analysis. However, for structures with high consequences of failure (such as high-rise buildings) or weather-exposed assets that lack a protective facade (such as bridges and open-air stadia), a continuous data stream from structural health monitoring (SHM) sensors can supplement and often extend the intervals between manual inspections. This approach can detect subtle damage earlier, leading to more efficient preventive maintenance.

The need for real-time insights has driven the development of digital twins—constantly updated virtual models that integrate a 3D BIM model with live sensor data. Typical measurements include acting loads, such as wind speed and direction or temperature, and structural response in the form of displacement, acceleration, and strain. The challenge then becomes how to interpret these heterogeneous data streams reliably and convert them into actionable maintenance decisions.

In the iconic Sydney Harbour Bridge for instance—a slight deviation from building structures, but SHM has a larger tradition in bridges—accelerometers are employed since 2014 (Costello 2023). The principle is straightforward: every structure has a characteristic vibration signature, and any physical change (e.g., a crack) subtly alters that signature. When a heavy vehicle such as a bus rolls over the deck it excites a very distinct response; if an arch is cracked, the resulting vibration pattern is measurably different. Because simple threshold alerts are unreliable, the Sydney Harbour Bridge uses a ML model: the one-class support vector machine (SVM) algorithm, which does not rely on neural networks but on predefined formulas. Each training sample is a fixed-length feature vector summarizing a brief window of vibration data from an accelerometer node. The SVM operates by identifying the best boundary (a hyperplane) to distinctly separate normal and abnormal data points, maximizing the margin between them. This algorithm is considered (semi-)unsupervised because the bridge authority lacks labeled examples for every potential damage scenario. Therefore, the model is trained solely on unlabeled "normal" data gathered during periods of known healthy operation. Consequently, the SVM learns the pattern of normal behavior without explicit damage labels, so new, out-of-distribution points recognized as abnormal. Finally, since 2014, the variety and number of sensors and algorithms used on the bridge have been continuously refined and updated.

Even though SVMs remain a common traditional ML choice for vibration-based SHM, researchers are now testing deeper architectures that offer extra flexibility (Zhang et al. 2022): one-dimensional (1D) convolutional neural networks (CNNs) digest raw acceleration windows, while two-dimensional (2D) and 3D CNNs mine image or spectrogram patterns; Long Short-Term Memory (LSTM) and other recurrent networks capture long-range temporal correlations in continuous streams; and autoencoders learn unsupervised representations for anomaly detection and data recovery.

Despite these advances, deploying deep learning on real bridges still encounters three primary challenges: a shortage of labeled damage data for training, (sensor noise and long-term drift that degrade the quality of data and the requirement for models that are interpretable and can be certified as safe for use in critical infrastructure. Nevertheless, some authorities are already testing such systems. The third-generation SHM upgrade on Hong Kong's Tsing Ma suspension bridge (Zhang et al. 2025) combines its existing fiber-optic network with AI-powered computer vision modules (high-definition cameras and weigh-in-motion sensors) to monitor traffic loads across the entire span in real time. This data is then fed into ML models that alert maintenance teams to any unusual structural responses.

Improving Point Cloud Workflows for Building Scans

This section also addresses the reduction of carbon footprints in the built environment by preserving structures. To renovate or extend the lifespan of an existing or historical building, it is essential to evaluate the building's structural capacity. Building scans are crucial in this process, as existing buildings often lack complete drawings and documentation, and construction deviations can lead to differences from the original plans. Therefore, these discrepancies must be considered in future planning. Furthermore, building scans not only aid in structural assessments but also help identify potential conflicts between new additions and the existing structure during renovations.

Point clouds represent the latest advancement in building scanning technology. They consist of a dense array of 3D coordinates accurately depicting a building's shape and surface. These are created by capturing numerous 3D points of an object or environment, typically using laser scanning (LiDAR) or photogrammetry. In the case of LiDAR, the process works by projecting laser beams and measuring the time taken to return, mapping distances; while photogrammetry involves using multiple images and algorithms to derive 3D coordinates from 2D images. Figure 6 illustrates a high-resolution Light Detection and Ranging (LiDAR)

Figure 6. Point cloud visualized with the online viewer usBIM.pointcloud.

point cloud that allows users to visualize and examine a historic building. These point clouds can also be imported into BIM software to assist in creating a BIM model based on the actual building's geometry.

ML and deep learning can greatly enhance the preprocessing, feature extraction, segmentation, and classification of point clouds compared to traditional methods, as demonstrated by companies like GeoAI, a surveying and mapping services provider (GeoAI n.d.). This improvement could therefore make sustainable renovation and rehabilitation projects more attractive to investors.

The need for these steps when handling point clouds arises because a raw point cloud from a building scan is a massive file, often several gigabytes in size, consisting of millions or even billions of unordered XYZ coordinate triplets. These are often accompanied by intensity; Red, Green, Blue (RGB) color; and scanner-pose metadata, all still tied to the local position of each scan head. Since the scanner is moved multiple times, the dataset usually contains dozens of overlapping, partially misaligned scan files. These files often include noise from reflective surfaces like glass, moving people or vehicles, stray atmospheric returns, and varying point densities.

Pre-processing raw point clouds is essential. First, individual scans must be stitched into a coherent coordinate system to ensure metric reliability, as even minor misalignments of a few millimeters can lead to centimeter-scale modeling errors. Following this, noise removal is necessary to filter out erroneous points caused by sensor inaccuracies or environmental factors. Next, downsampling reduces the number of points to make processing more efficient while preserving crucial information. Normalization then transforms the data into a common coordinate system or scale. Outlier removal eliminates points significantly different from the rest of the dataset.

After preprocessing, feature extraction identifies and calculates characteristics useful for distinguishing different parts of the point cloud. Segmentation follows, partitioning the point cloud into meaningful regions or objects. Finally, classification or modeling assigns labels to different segments or builds a model from the segmented data.

The deep learning approach begins by collecting training data, either using in-house labeled scans from LiDAR or photogrammetry, or from public datasets like ShapeNet and ScanObjectNN for objects, S3DIS and ScanNet for indoor environments, and KITTI or nuScenes for outdoor traffic scenarios. Most deep learning algorithms related to point clouds utilize supervised learning techniques. Some notable methods include the following: In terms of pre-processing, PointCNN performs an X-transformation to rearrange points into a standard layout before applying convolutional layers. Alternatively, VoxelNet and Sparse Voxel Networks start by voxelizing the raw point cloud, converting irregular points into a sparse 3D grid that is easier to filter and normalize. For feature extraction, PointNet applies a neural network to each point to extract individual point features and then merges these features using a symmetric function. For tasks like segmentation and classification, Stanford's PointNet++ employs a novel neural network structure similar to CNNs.

Typically, and also due to competitive reasons, private software companies do not disclose the specific algorithms and workflows they use for processing point clouds in their products. Nonetheless, users should be cautious about claims that AI-based software can automatically transform a LiDAR point cloud into clean floor plans or a detailed BIM model, as there is often a considerable amount of manual post-processing involved in these tasks.

Finally, AI-driven drones carrying lightweight LiDAR pods are now able to fly and map largely on their own, significantly speeding up the building scan process. They are equipped with onboard neural networks to detect obstacles, plan paths and steer the craft, while deep-learning-enhanced Simultaneous Localization and Mapping (SLAM) allow to localize the drone in real time, even where Global Positioning System (GPS) is absent. As the scanner fires thousands of laser shots per second, the AI dynamically adjusts altitude and speed to maintain uniform point density, closes coverage gaps and stitches successive sweeps into a geo-referenced point cloud that streams to the ground or is refined post-flight. The result is centimetre-accurate 3D data, captured in a fraction of the time and with significantly less human risk than manual piloting or tripod-based scanning.

Discovering New Construction Materials

Discovering construction materials that either result in lower carbon emissions during production or in superior mechanical properties for efficient material usage can greatly contribute in reducing the carbon footprint of new constructions or building expansions. ML offers significant potential in this process by enabling in silico exploration across a vast design space,[9] including not only theoretical crystal structures but also different cement compositions, alkali-activated binders, and mix designs incorporating industrial by-products or recycled aggregates. These digital evaluations prioritize candidates based on predicted strength, durability, and emissions from the process, effectively narrowing options before any physical tests are conducted. Nonetheless, the shortlisted formulations must undergo wet-lab validation,[10] which involves synthesizing powders or pastes at a bench scale, observing hydration or phase changes with X-ray and electron microscopy and assessing mechanical and durability characteristics before they can be used in construction projects.

The renowned AlphaFold AI system (Jumper et al. 2021) opened up this approach when it was successfully applied to the protein-folding puzzle, by treating it as a straightforward supervised learning task. It was trained using every high-quality structure from the Protein Data Bank, which consists of approximately 170,000 labeled pairs of amino-acid sequences and 3D coordinates. This dataset was expanded with millions of unlabeled sequences, which AlphaFold initially folded itself and then used as "pseudo-labels" in a self-distillation process that effectively quadrupled the data set. At the heart of AlphaFold's model is the Evoformer, a transformer stack that ingests multiple-sequence alignments and template features, learns long-range residue–residue relationships, and hands a latent "pair representation" to a Structure Module that iteratively outputs atomic coordinates. The entire network is trained end-to-end,[11] with the gradient derived directly from geometric losses like the Frame Aligned Point Error. The results, validated in the Critical Assessment of protein Structure Prediction (CASP) 14 blind benchmark,

[9] *In-silico* exploration simply means using computer simulations, models, and algorithms to explore, test, or predict material properties, behaviors, or reactions—without doing physical experiments in a lab. The term *in silico* comes from analogy: *in vivo* → in a living organism, *in vitro* → in a test tube/lab dish, *in silico* → in silicon (i.e., on a computer).

[10] "Wet lab testing" refers to experimental work done in physical laboratories that involves handling real materials, chemicals, or biological samples, often using liquids (hence "wet"), to measure, synthesize, or observe physical phenomena.

[11] In neural networks, "end-to-end" refers to training a single model to directly map raw input data to the desired output, bypassing the need for manual feature engineering or pre-processing steps.

achieved near-experimental accuracy, with a median Global Distance Test—Total Score (GDT_TS) of about 92, without relying on reinforcement learning or search heuristics. This demonstrated that a large, physics-aware supervised model combined with strategic self-distillation can solve a problem that had resisted dedicated wet-lab effort for decades.

This digital-to-physical loop approach to discover new materials is already being applied by Meta's data-center construction team. As part of their approach towards low carbon concrete in their datacenters (Meta 2024), working with the University of Illinois, Meta trained an optimization model built on its open-source Adaptive Experimentation (Ax) library. The model iteratively proposes concrete recipes, predicts the entire strength-versus-time curve, and scores each mix for embodied carbon intensity, cure time, and mechanical performance. In laboratory trials the AI-suggested formulations cut concrete-related emissions by up to 70 % in comparison to traditional concrete without sacrificing required strength, and Meta has already employed it at several US data center sites. Meta has open sourced the model and data so that other builders can tune the algorithm to their local materials, echoing the way AlphaFold weights are freely available to biologists.

A Bridge to Visual Programming for More Flexible Machine Learning Workflows in Building Design

Visual programming and more concretely the Grasshopper platform already mentioned in Chapter 3 is already a favorite among architects, engineers, and other building-design specialists for automating routine tasks. First, many of these professionals do not write code on a regular basis, so the ability to build scripts by visually wiring together components is immediately appealing. Second, Grasshopper ships with a rich set of geometry-oriented components that can be applied in seconds and whose results appear instantly in Rhino's 3-D viewport. These two factors make the idea of adding ML capabilities to Grasshopper—as a new family of components—an interesting idea for bringing designers closer to the possibilities offered by ML workflows.

There are several ML frameworks—used to code ML applications—that exist precisely because they bundle the tools, libraries, and GPU-level optimizations required to process large datasets and train complex neural networks efficiently. Among them, Google's TensorFlow and its high-level Keras API are two of the most widely used. Both, however, are written in Python, which makes them incompatible with the .NET-based environment of Grasshopper. To overcome this limitation, the open-source library TensorFlow.NET was used instead: an open-source adaptation in the C Sharp programming language that runs natively

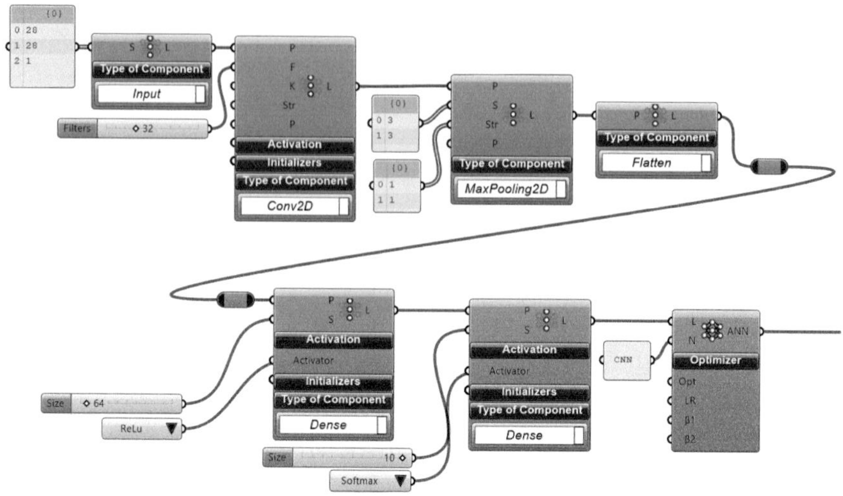

Figure 7. Definition of a convolutional neural network in Grasshopper using the plug-in Pug.

on the .NET framework. The result was the so-called plug-in "Pug" for Grasshopper, that implements several deep learning algorithms in the visual programming environment and already has several thousand downloads from the official food4rhino platform.[12] Figure 7, for example, shows how a CNN can be defined entirely within Grasshopper using the Pug components. With this visual-programming approach, not only is it straightforward to couple deep learning models to geometric problems defined in Grasshopper but experimenting with different network architectures or training hyperparameters also becomes far more approachable than in a pure code-based workflow.

The downside of this visual programming approach is that TensorFlow. NET, despite being a fantastic open-source effort, inevitably lags behind the Python-based TensorFlow. It cannot incorporate every new layer, optimizer, or research paper as quickly as the original library and it covers TensorFlow only, not the growing ecosystem of other ML frameworks like PyTorch, JAX, or ONNX. Consequently, some of the latest ML tools are not yet available inside the Pug plug-in. Even so, the workflows it does enable are already compelling. Figure 8 for instance shows the use of Pug training a neural network to predict the embodied-carbon footprint of truss structures represented as 3-D tensors (the preview is rendered with the third-party "Owl" plug-in). Such image-like encodings are handy

[12] The Pug plug-in is openly available on the Food4Rhino platform for anyone to download and try, and it is documented in Apellániz, Pettersson, and Gengnagel (2023).

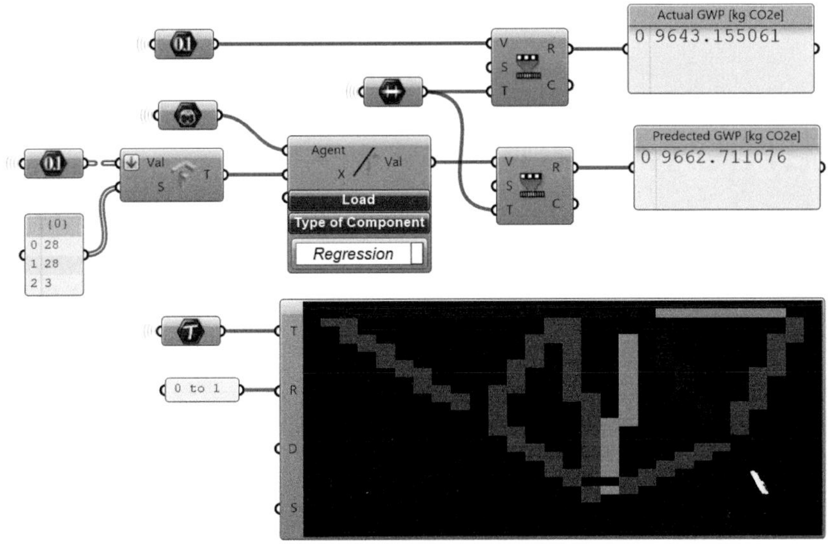

Figure 8. Prediction of the carbon footprint of truss structures with a supervised learning agent trained with the Grasshopper plug-in Pug.

later if those same structures are to be generated by other networks—for instance generative adversarial networks (GANs)—whose native output is precisely that kind of tensor.

Large Language Models

5

The release of OpenAI's ChatGPT to the public in November 2022 was a game-changer: for the first time, a powerful LLM was freely accessible through an ordinary web browser. Overnight, ML was no longer exclusive to tech giants and research labs, but it became a daily tool for everyone from journalists and architects to schoolchildren. Suddenly, unpromising model training processes were no longer required to apply ML for new use cases. It was as if those late-night TV "six-pack abs stimulators" finally delivered on their promise: instant results with zero sweat. However, this convenience came with concealed expenses. Behind every instant answer lies an energy-demanding infrastructure that cannot be overlooked. This chapter begins by exploring how LLMs function and are developed, providing a deeper understanding of their capabilities. It then features an interview with Alexander Forsch, a sustainability expert architect who has led the research and development (R&D) group at one of Europe's most technologically advanced architectural firms. Forsch offers insights into integrating applied research and LLMs in sustainable building design and dealing with the integration of energy-intensive data centers in sustainable urban development projects. The later sections try to squeeze the full potential of large language models for sustainable building structures. The chapter concludes with a reflection on how LLMs have empowered individuals with the new superpower of code vibing that can assist with daily tasks in building design. More specifically, this method can greatly aid in creating impactful LCA dashboards, thereby enhancing the collaborative sustainable building design process, as demonstrated by the findings of my doctoral thesis.

DOI: 10.1201/9781003506324-5

How Large Language Models Think

When ChatGPT[1] launched on November 30, 2022, and quickly gained a million users, it represented a public breakthrough for years of research. For the first time, anyone could interact with a neural network capable of writing prose, answering questions, debugging code, and even making jokes in real time, demonstrating that language generation had achieved a level of usability that previous AI had not.

This type of AI technology specializes in processing, understanding, and generating human language, and is known as an LLM, with the Generative Pre-trained Transformer (GPT) method becoming the norm since the introduction of ChatGPT. It is labeled generative because the model can create new text, sentences, or answers, rather than just selecting from a predetermined list like earlier ML classification models. It is pre-trained, meaning that before any interaction, the model has already been trained on a vast collection of books, articles, and web pages, eliminating the need for the user to provide additional training. The term Transformer refers to the specific neural-network design it utilizes, which excels at recognizing how words in a sentence relate to each other (it "pays attention").

Figure 1 illustrates an LLM named "nanoGPT",[2] which comprises three stacked transformer blocks that execute the mathematical processes from an input prompt to the resulting text. This instance of nanoGPT contains only 85,584 parameters, whereas modern state-of-the-art models boast tens to hundreds of billions, an increase of about six orders of magnitude. With enough data and computational power, increasing the number of parameters tends to enhance the quality of language models significantly. Why this sets transformers—and the LLMs built on them—apart from earlier deep-learning architectures is because their self-attention mechanism models long-range dependencies and maps efficiently onto GPUs. As a result, scaling up parameters and data delivers larger quality gains than comparable scaling of dense or convolutional networks. It is as if these earlier networks are trying to read a novel through a keyhole: they only glimpse a few words at a time, so adding pages (data) doesn't help much. A transformer opens the whole book on a table. More pages now provide genuinely new information, and a bigger magnifying glass (more

---

[1] ChatGPT is OpenAI's conversational interface—available via web, mobile apps, and an API—powered by the company's proprietary GPT family of large language models.

[2] nanoGPT is a concise, open-source PyTorch implementation of a GPT-style Transformer. With only a few hundred lines of code, it demonstrates the core training loop and model architecture of large language models, making it a popular educational reference and sandbox for rapid experimentation (Karpathy 2023).

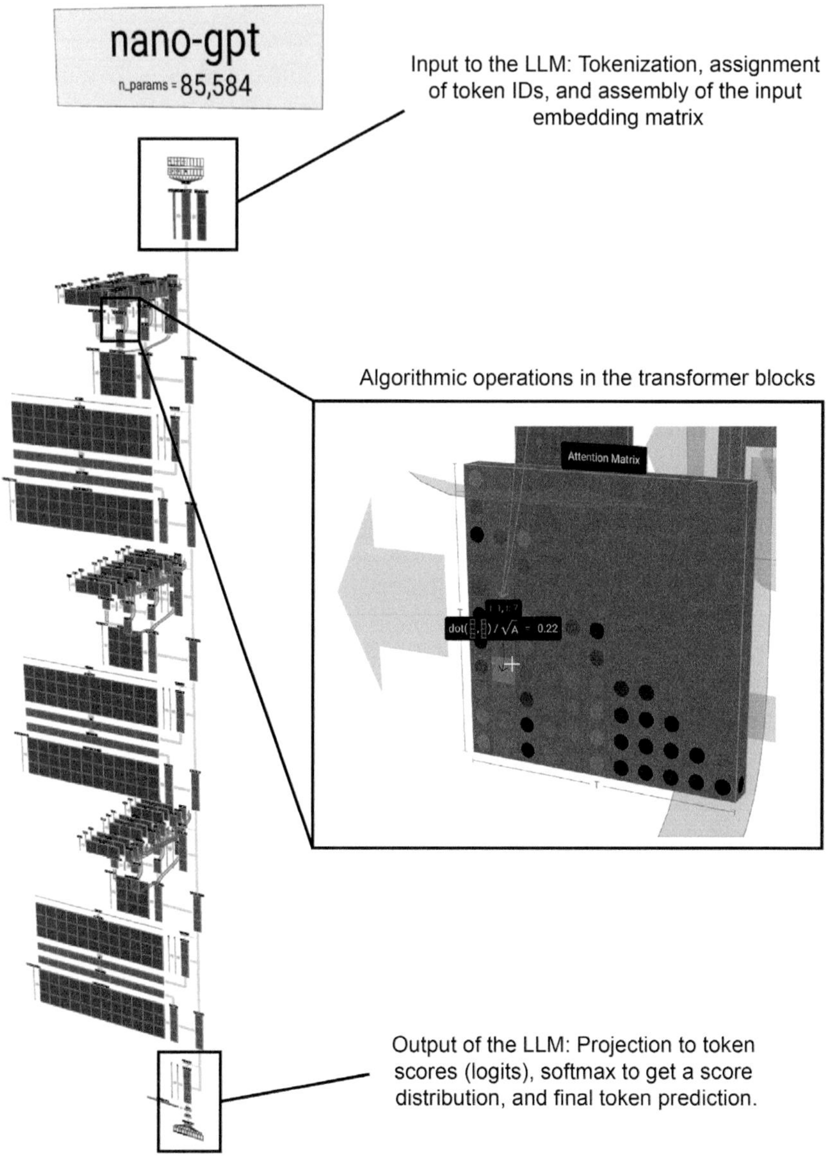

Figure 1. Visualization of an LLM with the online viewer from https://bbycroft.net/llm.

parameters) lets the model notice subtler details, while the GPU acts like an efficient page-scanner that can process everything in parallel.

To grasp how LLMs "think," it is essential to understand their method of converting text into numerical form. As discussed in the previous chapter, neural networks can only process numbers, so characters or words must be transformed into numeric vectors. One-hot encoding, previously mentioned, depicts each word as a large sparse vector. However, modern LLMs adopt a more efficient approach: they break down the input string into a sequence of tokens (subwords or word-pieces). Each token is then looked up in a vocabulary table and replaced by its token ID, a single integer.

These token IDs are not suitable for heavy algebraic operations yet, so the model subsequently performs a memory lookup into a learned embedding matrix to find a so-called embedding vector associated with that token ID and replace it. The output, which serves as the input for the first transformer block, is a matrix with dimensions (sequence_length × embedding_dim), prepared for the transformer's multi-headed self-attention and feed-forward layers. Within each transformer block, the network performs hundreds of millions of dot products, additions, and non-linear activations, gradually mixing information across token positions and building up contextual representations.

After the transformer layers process the input, the resulting hidden vectors are converted back into token space. This is done by multiplying each vector with the transpose of the embedding matrix, or alternatively, with a separate output weight matrix. This multiplication results in an output vector that matches the size of the token vocabulary, where each element corresponds to a score for each possible token. From these scores, the language model either selects the token with the highest score, akin to ML classification methods, or randomly chooses a token based on the score distribution. The model's temperature setting allows the user to control this randomness. The selected token ID then forms the next set of characters in the output sequence. By continuously repeating the cycle of tokenize → embed → transform → decode, the LLM produces a sequence of tokens. When these tokens are converted back into their corresponding subwords or word-pieces, they create coherent text, provided the model has been appropriately trained.

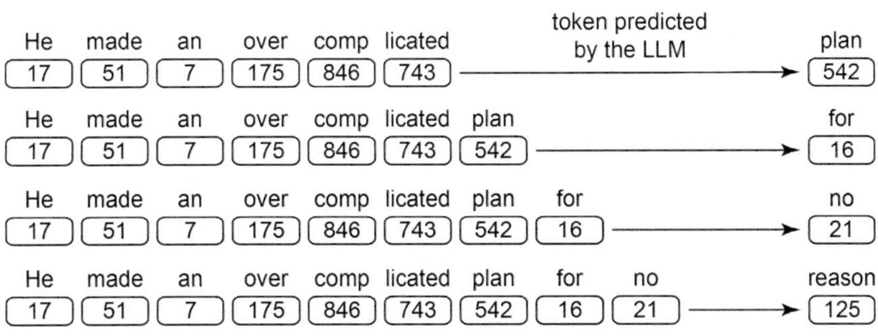

Figure 2. LLM Inference: Generating Predictions.

Figure 2 illustrates an LLM in action. The model constructs text incrementally, one token at a time. Initially, the user's prompt is converted into integer IDs through tokenization. At every step, the LLM selects a token ID, which is then added to the next available position in the sequence. This updated sequence is re-entered into the model, and the process continues until a special (end-of-sequence) token is encountered. This phase is called inference because the network is generating—or inferring—new text, rather than replicating strings from its training data.

The last aspect to grasp about how LLMs operate is their training process. Firstly, they must learn to generate coherent text. Secondly, they need to be capable of interacting with users like ChatGPT and similar platforms, rather than merely predicting the next token, as shown in the previous illustration.

Pre-training is the first step, involving the creation of a foundational model, like the aforementioned "nanoGPT," which is focused solely on generating text predictions. To achieve this, LLMs are trained on an extensive, unfiltered text corpus (including web pages, books, articles, code, etc.) using self-supervised learning. This process involves predicting missing or subsequent tokens based on the surrounding text, essentially filling in gaps or anticipating the next word. By completing millions or even billions of these small prediction tasks, the network develops a deep understanding of grammar, facts, and common patterns. At this point, LLMs can generate coherent text, which may not even be based on real facts, but they do not know how to properly reply to a human user prompting questions.

Once the LLM has processed a sufficient amount of text, it moves on to the supervised fine-tuning stage phase, where it learns to follow instructions rather than just generate coherent text. A collection of prompts is paired with human-created examples of ideal responses. These sample answers cover various tasks and styles, including translation, summarization, and question answering. The pre-trained model is then fine-tuned[3] using these prompt-response pairs, allowing it to learn through supervised learning to replicate the given outputs when presented with the same prompts. This phase solidifies the model's extensive language comprehension into specific instruction-following capabilities.

---

[3] Fine-tuning is the process of taking an LLM that has already been pre-trained on vast, general-purpose text and further training it—using supervised learning—on a curated set of prompt–response pairs. Unlike full retraining, which requires training a model from scratch, fine-tuning starts from an existing model and adjusts its parameters to specialize in producing desired outputs (such as translations, summaries, or answers) when given similar prompts. Because it usually updates only a subset of the model's parameters or focuses on specific tasks, fine-tuning is significantly less computationally expensive and requires far less training data.

Even though the model has been trained to follow instructions via supervised fine-tuning, it still does not fully align with human preferences. The final two training steps aim to address this. First, in the reward-model training phase, a separate ML model is trained via supervised learning to predict human preference rankings among multiple response options for a given prompt. The training data consists of responses generated by the fine-tuned LLM for each prompt, which human annotators then rank from best to worst.

In the final training stage known as "reinforcement learning from human feedback" (RLHF), the LLM refines its behavior by using the trained reward model within a sample–score–update loop: for each prompt, the model generates a batch of candidate responses, the reward model assigns each a score, and the LLM's parameters are updated via reinforcement learning so it more likely generates higher-scoring outputs. This automated loop removes the need for human annotators to rank outputs during training. Over time, the model learns to prefer responses that better align with human evaluations, yielding outputs that are safer, more useful, and more accurate.

The LLM is now prepared for public deployment. Yet several challenges still persist. Firstly, there are numerous ethical concerns about the training process of LLMs that will be discussed in a different chapter. Secondly, if the model is to be used in relation to domain-specific or private data, publicly available LLMs might not have had access to such information during training. This chapter will explore different approaches to tackle this challenge in order to apply LLMs to the specialized field of sustainable building structures.

Conversation with Alexander Forsch About Leveraging Applied Research and Large Language Models to Enhance Project Sustainability and Strategies for Managing Data Centers Within Sustainable Urban Planning

DA: You have had a career at several leading architectural firms, and now you're joining Witteveen+Bos. Let's revisit when we first met. Back then, you were Senior Environmental Designer at MVRDV architects, leading the climate segment of the R&D team known as "MVRDV Next". This was also an exciting period with the rise of generative AI. Can you give us a brief overview of your career journey and explain how you transitioned from being an architect project lead to leading a sustainability-focused R&D group? What advantages did you find in this transition compared to your colleagues who worked solely in R&D?

AF: My academic background is in architecture and urbanism, with a strong focus on sustainability. After starting at MVRDV as a project architect and advancing to a design lead, I realized my desire to focus more on sustainability. To achieve this, I pursued a postgraduate program at ETH Zürich, which was quite challenging as I was working at the same time. This program explored topics such as embodied carbon, alternative construction methods, and LCA. Subsequently, I transitioned to the R&D team at MVRDV. This team, known as MVRDV Next, was initially established a few years earlier with a focus on algorithmic and computational design. As the company sought to enhance its sustainability efforts, a climate-focused segment was created, with me leading it. The goal was to merge sustainability with computational processes. In this role, we developed tools like CarbonScape for early-stage carbon monitoring, which was instrumental in guiding our design teams' decisions and communicating them to clients, demonstrating that these decisions are not just driven by personal interest but have measurable impacts on carbon footprints. My previous practical experience provided a unique perspective in the R&D group, allowing me to understand the timing of project decisions and the necessity of our input. While there is extensive research and specialized knowledge, integrating it into actual projects is a different challenge. Additionally, with the rise of generative AI, colleagues like Cas Esbach and Freddy Forsyth contributed to researching how to integrate AI tools into our workflows. This collaboration proved to be a successful synergy, combining sustainability, computational design, and AI effectively.

DA: You were involved in a certain project focused on developing an office campus intended to accommodate several AI companies. This project included designing numerous buildings with varying functions, sizes, scales, and requirements. Additionally, the client requested an on-site data center to support the AI workloads while still aspiring to achieve a top-tier green certification. Considering the substantial energy and cooling needs of a data center, how did you address this challenge and ensure the project remained on course for certification?

AF: From the start, we integrated the data center into the campus's overall energy strategy, rather than treating it as an isolated unit. Our climate team participated in every coordination meeting, alongside structural engineers and energy consultants, to ensure its design aligned with the masterplan. We strategically positioned the data center at the center of the campus, enabling its waste heat to be directly utilized in the campus district-heating loop Additionally, we collaborated with the municipality to explore broader applications, such as thermal storage and heating at the neighborhood level, turning what is usually a by-product into a valuable campus resource. We also aimed to reduce the embodied carbon in the facades of the campus. Was triple glazing necessary if hundreds of kilowatts of heat were being pumped in? In some areas, we switched to double glazing to reduce embodied carbon without affecting the overall energy balance. This was my first experience with data-center planning, and understanding the extensive electricity and water needs, especially in discussions with the city council about water-use restrictions, highlighted that these buildings are more than just "IT rooms". They are energy infrastructures. Ultimately, the client opted for some cost-effective configurations for the data center, balancing budget considerations with performance, but we still maintained our holistic, cross-disciplinary approach in the design of the campus to achieve green certification. As for whether AI's energy consumption is justified, decarbonization is an urgent challenge for our generation, and AI can be a powerful tool for that. However, political will is essential to transform the energy sector towards renewables, smarter grids, and scalable storage, ensuring that AI's energy needs are met with clean power, leading to real emission reductions and sustaining further innovation.

DA: As part of "MVRDV Next" in aiding the office's design teams, you played a role in creating the internal "carbon guidelines".[4] Could you elaborate on the inspiration behind this idea and how these guidelines were implemented in the campus project?

AF: Our goal with these carbon guidelines was to simplify the concept of embodied carbon for architects and designers, making it easier to grasp without exploring LCA calculations in depth. We wanted architects to develop an intuitive understanding of embodied carbon, similar to how they develop a feeling for cost at some point. Using these general carbon guidelines, we collaborated with various stakeholders and teams to create what we call material diets for the different building projects on the campus. Each project presents unique challenges in finding sustainable solutions. For smaller buildings, we could almost entirely utilize mass-timber construction. For mid-rise office buildings, we suggested hybrid timber structures—combining mass timber with concrete—where all-timber designs were impractical due to spans, exposure, or building

[4] The carbon guidelines from MVRDV (2023b) are presented as a poster that lays out core strategies for minimizing embodied carbon in early design stages, including: integrating existing site materials and structures; avoiding new and underground construction; designing for future flexibility and programmatic synergies (e.g., shared facilities, flexible rooms, activated transitional spaces); rethinking parking to reduce demand; reducing structural weight and optimizing floors; simplifying facades and openings; and planning for demountable, bio-based, or recycled components. It highlights that structure, floors, and facades account for most embodied carbon, urging focused LCA on stages A1–A3 and the tailored combination of guidelines per project typology.

codes. This project-by-project guided approach maintained flexibility, ensuring each campus building achieved its sustainability goals and enabled designers to make informed, carbon-conscious decisions from the start. Additionally, a crucial challenge was identifying the availability of sustainable or alternative building materials, determining which companies produce them, and how close these are to the construction site.

DA: Apart from the campus project, your carbon guidelines stress the importance of maintaining existing buildings whenever possible. What are the primary obstacles to reusing buildings, and why do we continue to witness a significant amount of demolition and new construction? How do you address the challenge of evaluating an existing building when there is limited documentation available?

AF: The primary focus should be on preserving building structures. Meeting climate targets is unattainable if we continue demolishing and constructing at our current rate, even if the new buildings are green or sustainable. Therefore, it's essential for architects to emphasize this in the initial design stages. Conducting preliminary calculations or using ML-based carbon predictions early on can highlight the advantages of maintaining existing structures or making slight modifications or extensions. This approach can significantly influence a project's outcome. AI can assist architects in presenting arguments to support decisions in favor of preserving existing structures. Regarding the documentation challenge, from my experience, it's beneficial to perform a rough scan at the outset. With specific knowledge about existing buildings, if a building was constructed in a certain era, it might contain asbestos that must be removed before renovation, and it is crucial to identify this early. Additionally, knowing if there were significant renovations during that period can indicate potential complications in the renovation project. You can also make informed guesses about construction methods, such as whether it was masonry or the type of slabs used, based on the building's age. This insight can help facilitate discussions with clients at the project's inception and ensure it starts on the right path.

DA: As we wrap up the interview, I can envision using LLMs for several topics we have covered: offering architects guidance based on the carbon guidelines, assessing an existing structure for renovation possibilities based on its location and construction date, and providing details about sustainable materials for specific locations, for instance. Do you incorporate ChatGPT or other LLMs in your daily work routine? Since you are a DGNB certified manager, you occasionally need to draft reports to secure green building certifications for your projects from this organization. Are you employing an LLM for these comprehensive reports, or is it primarily still done manually?

AF: Generative AI, particularly LLMs, can significantly accelerate certain processes. However, I prefer using them to create the final document, and human involvement is still essential not just for providing prompts but also for checking and adjusting the output. AI's usefulness depends on the user's knowledge of the subject, which ensures control over the outcome. The better trained the person is in their field, the better the results will be with AI.

DA: Ok so it is evident that AI can assist us, but it is far from taking our jobs let's say.

AF: That is my point of view… at the time of this interview.

AI Bots for Sustainable Building Design

The previous interview revealed that there are numerous subjects regarding sustainable building design where architects and engineers could use support or guidance from AI. Some of those topics might fall outside the scope of conventional architectural or engineering education, such as carbon guidelines and the availability of alternative building materials. Others might be too vast to remember, like the history of construction related to a specific area or past renovations. Additionally, tasks like writing detailed documents according to specific guidelines or formatting rules, such as sustainability reports for green building certification, could just benefit from increased productivity.

AI bots can be used in the building design process to address these challenges. They are software programs created for textual or spoken interactions that offer numerous inherent advantages: they are always available, can scale to manage multiple tasks or users at once, can quickly access extensive information, and have the potential to learn from each interaction, thereby becoming more accurate and tailored over time. Consequently, they can assist with sustainability-related tasks, allowing designers to concentrate on higher-level creativity and strategic decisions while still achieving sustainable building designs. The primary challenges in utilizing AI bots for sustainable design include finding an LLM with enough reasoning capability to meet designers' needs; ensuring that chatbots are equipped with the necessary specialized knowledge of sustainable buildings; and selecting a chatbot service to host the LLM that ensures data privacy, integration flexibility, cost-effectiveness, quick response times, and robust community support.

The range of options for selecting an appropriate LLM and an LLM-powered service through simple web interfaces has expanded significantly since OpenAI brought conversational AI into the mainstream with ChatGPT, as discussed by Andrej Karpathy.[5] ChatGPT serves as a chatbot service, with the underlying LLMs being OpenAI's GPT model family, including the latest GPT-5 as well as the previous GPT-4o, GPT-o4-mini, and others. Other major tech companies provide similar models and/or services—for instance, Google's Gemini model (accessible to end-users through the Gemini chat interface), Meta's Llama family of models (available in the Meta AI assistant), and Microsoft's Copilot service, which primarily uses OpenAI's GPT-4-series models. Additionally, several start-ups offer notable LLM services: Anthropic's Claude (based on the Claude-3 model family), xAI's Grok (built on Grok-1), Perplexity AI's answer engine (which combines multiple third-party LLMs), DeepSeek's DeepSeek-V2 model, and Mistral's Le Chat interface (powered by the Mistral Large and Mixtral models). All these services are provided as web-based chatbots; signing up usually requires just a few clicks, and most offer free or trial tiers, making interaction with advanced LLMs—as shown in Figure 3—highly accessible.

Beyond proprietary LLMs, where the model architecture and weights are kept hidden, there is a thriving ecosystem of open-source LLMs available for download from platforms like Hugging Face. An LLM is genuinely "open source" only when both its code and parameters (model weights) are released under a permissive license, allowing users to run

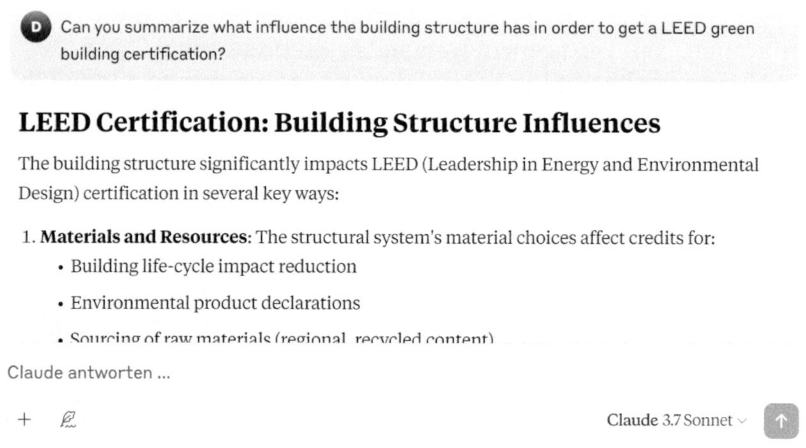

Figure 3. LLM Claude 3.7 Sonnet accessible on free tier at claude.ai.

[5] Karpathi (2025a) presents a practical walkthrough of LLMs and their growing list of related features, as a new entry to Karpathy's general audience series on LLMs. Karpathy is a Slovak-Canadian computer scientist, founder of Eureka Labs, former Director of AI and Autopilot Vision at Tesla, and a founding research scientist at OpenAI.

it locally—that is, perform inference on an own (central processing unit) CPU or GPU without transmitting data over the internet. Running models locally offers clear privacy benefits, as your prompts and data remain on your machine, which can be vital for confidential building design projects. However, the downside is the hardware cost: replicating the quality of a 30B+ parameter model typically requires one or more high-end GPUs, such as NVIDIA A100/H100 with 40–80 GB Video Random Access Memory (VRAM), costing thousands of dollars. Thus, most users opt for smaller (4–7B parameter) and less capable versions to stay within the limits of a single-GPU setup. To experiment without downloading multiple GB-heavy models for each test, services like app.hyperbolic.xyz offer pay-as-you-go inference on dozens of open-source LLMs. When it comes to architectural or engineering offices using local LLMs, it is more energy- and cost-effective, as well as equally secure in terms of data privacy, to deploy a single local LLM on a dedicated virtual machine accessible on the office network. This way, all employees can use the same instance rather than installing individual copies on each person's computer.

The most reliable way for building designers to pick a chatbot service (for example, ChatGPT) and select the best LLM it offers (such as GPT-4o) is simply to try them out with real questions. A basic test might be, "What different kinds of slab constructions are there?" A correct response is likely because timber construction is a well-documented, longstanding topic in the public web data on which LLMs are trained and it is not affected by recent cut-off dates (information made public after the model was trained). A tougher test could be: "Which slab constructions would I expect in a 1932 building in city ABC that is now being renovated?" That requires more narrowly focused historical or regional data. Finally, an even more demanding prompt to help choose an LLM would be: "Given a specific occupancy load and column spacing, how thick should my CLT slab be?" Correctly solving that involves not only knowing relevant building codes (if they were in the training set) but also carrying out multi-step reasoning.

To improve the odds of a valid response, it is encouraged to use prompt-engineering[6] techniques, especially chain-of-thought prompting (for instance, "Please explain your reasoning before giving the numeric answer"), since LLMs generate text one token at a time and often perform better when the logical steps come first. Therefore, in case well-justified and deductive answers are regularly needed, it is advisable to look for a so-called reasoning model that internally runs a structured inference process

[6] Prompt engineering is the practice of designing and refining natural-language prompts to guide generative AI models toward producing the most accurate, relevant, and coherent outputs. It encompasses choosing precise phrasing, supplying contextual information or examples, and specifying stylistic or task-oriented instructions to optimize model behavior.

even if the user does not explicitly request it. Still, no model is infallible: LLMs can "hallucinate", offering plausible sounding but fabricated details (e.g. "the wall should be two meters thick") instead of sound engineering judgment. In structural or architectural design, precision and accuracy are nonnegotiable, so models known for strong inference capabilities and low hallucination rates should be the norm. Hallucinations can also happen, not due to a failed reasoning process, but simply because the information needed was not in the training corpus. A practical way to detect possible hallucinations is to query several different LLMs and compare their answers—thus, creating a so-called council of LLMs.

Although users might feel overwhelmed by the wide range of chatbot services and models available, platforms like lmarena.ai are designed to help by continuously ranking these LLM versions. This is done through human judges who blindly vote on which chatbot response is better, as illustrated in Figure 4. Additionally, they provide a link to the hosting service, allowing users to test the model themselves—which is highly recommended—and explore and compare the well documented features of each chatbot service. However, there is currently no public "Green Arena" to rank models based on their carbon footprint or energy efficiency, as providers seldom disclose power metrics per query. Therefore, designers should choose the LLM and chatbot service they are most comfortable with, as the improvements in optimized building performance will more than compensate for the AI's energy consumption (see Chapter 1). And, naturally, using AI for personal productivity or even leisure remains perfectly acceptable—so long as each of us exercises it responsibly.

Nevertheless, the previous prompt examples assume that the information required to generate a response is already included in the LLM's pre-training data. However, this is not always the case. For example, if a structural engineer queries "We want to explore using inner bearing walls made of clay masonry instead of carbon-intensive concrete

Rank* (UB)	Rank (StyleCtrl)	Model	Arena Score	95% CI	Votes	Organization	License
1	1	Gemini-2.5-Pro-Preview-05-06	1446	+8/-9	4500	Google	Proprietary
2	1	o3-2025-04-16	1413	+8/-7	6689	OpenAI	Proprietary
2	3	ChatGPT-4o-latest (2025-03-26)	1408	+6/-6	10290	OpenAI	Proprietary
3	5	Grok-3-Preview-02-24	1403	+4/-4	14843	xAI	Proprietary
4	3	GPT-4.5-Preview	1398	+4/-5	15275	OpenAI	Proprietary
4	5	Gemini-2.5-Flash-Preview-04-17	1394	+7/-7	5959	Google	Proprietary
7	5	DeepSeek-V3-0324	1373	+7/-5	8753	DeepSeek	MIT
7	5	GPT-4.1-2025-04-14	1366	+7/-8	5102	OpenAI	Proprietary
8	8	DeepSeek-R1	1358	+4/-4	18493	DeepSeek	MIT

Figure 4. Top-ranked LLMs as listed on lmarena.ai, accessed 12 May 2025.

walls; what thickness should they be given our floor-to-ceiling-height requirement?" and if the pertinent regulation has just been issued and is not publicly available, the model does not have that knowledge and needs to be provided with that text during inference. The most straightforward approach is contextual prompting, also referred to as cache augmented generation (CAG), that would mean adding the entire regulation to the user prompt, as long as it fits within the model's context window. The limits of context windows have expanded significantly: the GPT-3.5 model, which powered the initial ChatGPT release, accommodated 4,096 tokens; GPT-4o now supports 128,000 tokens; and Google's Gemini family offers 1–2 million tokens for models requiring extensive context. LLM organizations clearly recognize the significance of context length in delivering domain-specific AI tools, which is why they have been vigorously researching LLM architectures with these capabilities, as illustrated by Google's recent research paper "Leave No Context Behind: Efficient Infinite Context Transformers with Infini-attention".[7]

A more effective method for delivering extensive, unchanging context to an LLM rather than pasting it in user prompts, is through the system prompt. This system prompt serves as the initial instruction given to an LLM, typically marked with the role system, and it establishes the model's overall behavior, tone, and limits for the remainder of the interaction. Internally, system, user, and assistant messages are combined and tokenized, with models specifically trained—via RLHF—to follow these initial guidelines. Figure 5 illustrates how information is automatically converted into tokens by LLMs behind the scenes, enabling these models to recognize and differentiate between different message roles. In the case of OpenAI's ChatGPT, it provides "custom instructions" for individual conversations and a "custom GPTs" feature that allows users to create and publicly share their own instructed bots in the "GPT Store" or keep them private within an organization. However, not every chatbot service offers a user-adjustable system prompt. Beyond user interfaces, most LLM APIs (such as those from OpenAI, Anthropic, Google, etc.) permit developers to add custom system messages—allowing teams to create and deploy their own specialized chatbots through these APIs.

However, a significant challenge persists when the contextual information required to direct a sustainability-focused AI bot surpasses the LLM's maximum context window. Also possible, it might not be efficient

[7] Munkhdalai et al. (2024) introduce Infini-attention, a novel mechanism that embeds a compressive memory alongside both masked local attention and long-term linear attention in each Transformer block, enabling LLMs to process virtually infinite context lengths with bounded compute and memory requirements.

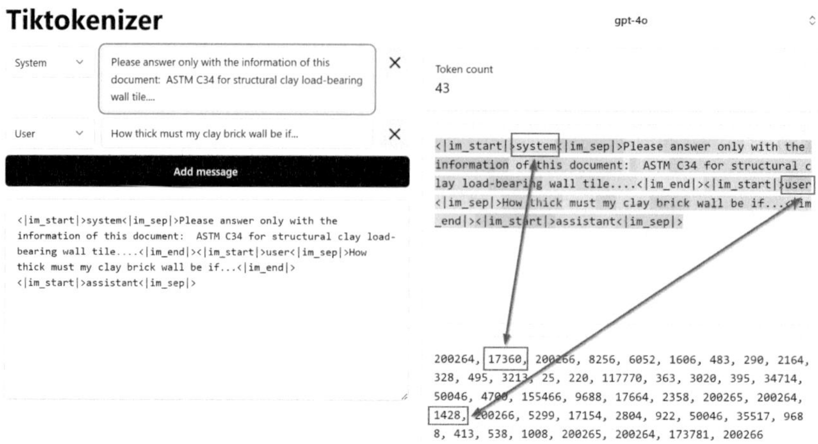

Figure 5. Visualization of the translation of system and user prompts into token IDs at tiktokenizer.vercel.app.

to supply excessive contextual information when only small fragments are needed for an accurate response, as this can considerably slow down the model's performance and increase operating costs.[8] Retrieval-augmented generation (RAG) and fine-tuning are approaches that address this issue, as detailed in a subsequent section. Furthermore, this chapter suggests that LLMs are useful only for specific portions of the design phase, mainly as chatbots (as the infamous giant ad wrapped around an unfinished building in Antwerp in June 2023 that begged, "Hey, ChatGPT, finish this building ..." manifested). However, this apparent limitation vanished with the arrival of AI agents, which will be discussed in Chapter 7.

Embeddings and Semantic Searches in Life-Cycle Assessments

Embeddings are numerical vectors that represent data—such as text, images, or audio—in a way that preserves the semantic relationships between items. While they play a crucial role in the RAG approach detailed in the following section, they can also effectively function as independent search algorithms, particularly in processes dealing with unstructured data, such as LCAs.

[8] The length of the input sequence directly impacts the computational resources required by an LLM, primarily due to the quadratic scaling of the attention mechanism and the increased memory needs for storing activations.

Embeddings have gained significant importance in today's deep learning frameworks.[9] They were first introduced in this book when it was explained that all ML models require numerical input data and that LLMs convert the integer token IDs of an input text sequence into a numeric embedding matrix—one embedding vector per token ID—which is then fed through the transformer layers. In addition, there are standalone embedding services that map various kinds of data (text, audio, images, etc.) into fixed-length numeric vectors—such as OpenAI's "text-embedding-ada-002" model (text embeddings) or their open-source Contrastive Language–Image Pretraining (CLIP) model (text and image embeddings). For instance, the Ada embedding model generates vectors with 1536 dimensions. Although the individual dimensions of these embeddings are not easily interpretable by humans, similar to the parameters of most deep learning models, the key point is that vectors located near each other in this 1536-dimensional space typically represent inputs with similar meanings. A common method for measuring this is cosine similarity, which calculates the cosine of the angle between two vectors, with values ranging from "–1" (completely opposite) to "+1" (identical).

Semantic search using embedding vectors is utilized in technologies such as recommendation systems (for example, Netflix can suggest shows similar to the ones users already watched) and search engines (like how Google finds pages whose content embeddings closely align with the search query). The major benefits of semantic search are that the data does not need to be sorted into strict categories, and that the search text does not have to precisely match any entry. In these mainstream technologies, in order to handle similarity queries over millions of high-dimensional vectors, "vector database" structures are used, that support efficient inserting, deleting, updating, and rapid similarity searches. With the growing popularity of LLMs and RAG, more and more providers are now offering managed vector databases and semantic-search services.

Figure 6 demonstrates how semantic search can be applied to an LCA. When mapping a building's inventory, there are thousands of EPDs for construction materials to choose from. However, finding the specific declaration needed can be challenging due to the lack of a standardized naming convention for EPDs and the overly broad categories in databases. In this example, even if the term "timber" is inputted with a spelling

9 Embeddings rose to prominence in modern deep learning after Google researchers released word2vec in late 2013, demonstrating a fast, scalable method for learning high-quality word vectors and sparking widespread adoption across natural language processing (NLP) and other data modalities (Mikolov et al. 2013).

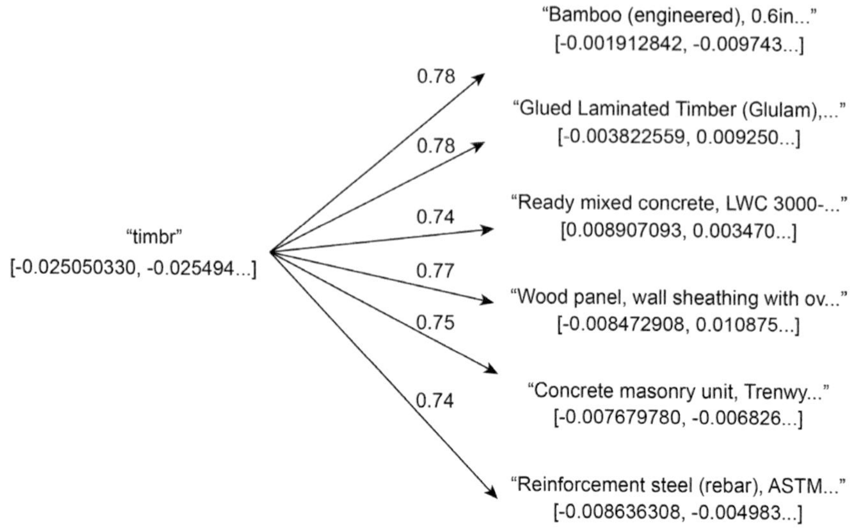

Figure 6. Semantic search applied to LCA.

error, it still retrieves items related to wood products that have high cosine-similarity scores, despite none of the names including "timber." Instead of focusing on exact cosine results, most semantic-search systems rank candidates based on similarity and provide the top-n results, which typically suffices for the user.

For instance, the commercial LCA software for buildings One Click LCA includes an "AI mapping" feature, illustrated in Figure 7, designed to automatically align imported data from various 3D modeling software, like Revit family names, with an EPD from its database. However, for competitive reasons, the software does not disclose which model is utilized for these mappings or whether it uses embedding semantic search.

Expanding Sustainability Bots' Knowledge Through Retrieval-Augmented Generation and Fine-Tuning

The sustainability AI bots introduced earlier faced a limitation in their ability to handle private or recent data, as this information was not used to train the powerful publicly available LLMs. To augment the context that the LLM can process beyond its context window, there are two fundamentally distinct approaches: RAG and fine-tuning.

RAG is an algorithm or technique in which a user's query to an LLM is enriched with context snippets drawn from an external, privately held

Name in imported data	Map resource Click a name to change (single row)	Mapping basis
Concreto C35/45	Concrete, C35/45 (B35 M40) (ex rebar) (..	AI mapping
rebar	Concrete, C50/60 (ex rebar) (IZB)	AI mapping
Brick wall earth bricks	Brick wall, incl. mortar, with Okobau.dat ..	AI mapping
Concreto C30/37	Concrete, C30/37 (B30 M60) (ex rebar) (..	AI mapping
rebar	Concrete, C50/60 (ex rebar) (IZB)	AI mapping
Window triple glazing	Window with insulated triple glazing, per ..	AI mapping

Figure 7. AI mappings in the software One Click LCA.

database.[10] Initially, lengthy documents are divided into smaller sections or "chunks." Each chunk is converted into an embedding vector and stored in a vector database. When a user submits a query, the query is also embedded, and a retriever identifies the most semantically similar chunks. This search process can be enhanced with keyword matching or by categorizing entries. Only the most relevant chunks are then appended to the prompt and fed into the LLM. Thus, the LLM itself does not require any internal modifications beyond a simple instruction to "answer based only on the context below" aimed to prevent hallucinations, assuming the model has adequate inference capabilities. Furthermore, including source labels in each chunk allows the LLM to cite the origin of its information. Finally, since all knowledge resides in the database, updates or expansions are straightforward, eliminating the need for costly model retraining and ensuring the LLM always reflects the most current data, whether private or public.

Figure 8 shows a chatbot powered by a RAG system that aids structural engineers in navigating the complexities of retrofitting and renovation projects. These kinds of projects help to reduce carbon footprints compared to the "demolish and rebuild" method. Therefore, this tool enables engineers to better assist building developers with the intricate decision-making process for existing buildings and also decreases planning time and costs, encouraging developers to choose renovation. The idea is to grant the chatbot with access to all previous office projects, but it is a dataset too large to fit into a language model's context window simultaneously. Instead, the system retrieves only the most pertinent

[10] RAG was introduced in 2020 by Patrick Lewis and colleagues at Facebook AI Research. Their NeurIPS paper (Lewis et al. 2020) proposed coupling a sequence-to-sequence language model with a neural retriever over an external Wikipedia index, demonstrating large accuracy gains on knowledge-intensive tasks and establishing the architecture now widely referred to simply as "RAG".

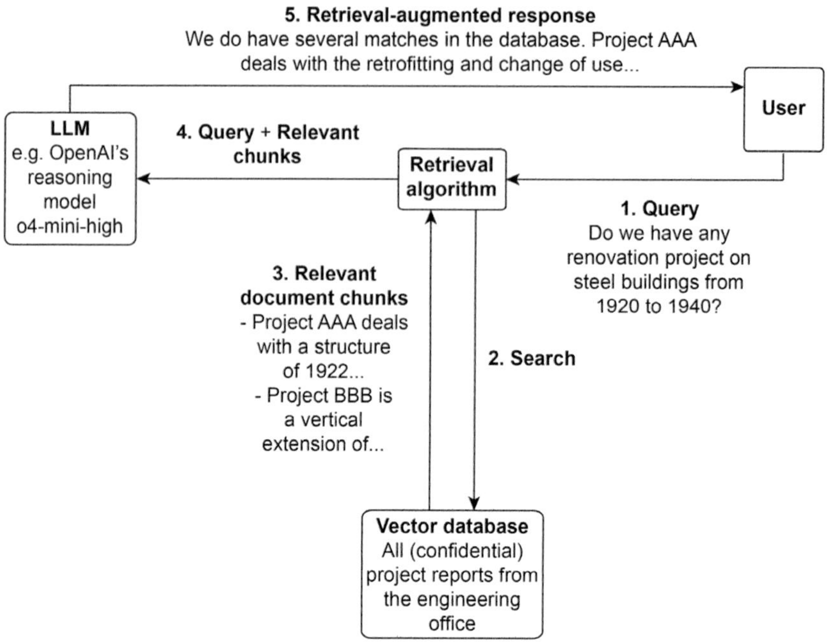

Figure 8. RAG chatbot for building reuse and retrofitting.

project records at the time of a query. Thus, engineers can inquire about design decisions in past projects to assist their decision-making for new projects, and the LLM provides responses tailored to their questions much faster than manually searching through a project server.

There are various ways to set up RAG workflows, but users must find a balance between usability, the control of the algorithm, and the privacy that each method provides (since the primary goal of RAG is to link a chatbot with private data). For example, the engineering company in the previous scenario would not want to keep confidential project reports on a server without strong security measures, nor would they want those reports to be used to train an LLM that might later expose them to other users. The simplest options for RAG are the built-in features in many modern chat interfaces, which do not require any coding. For instance, ChatGPT has an "Upload Files" feature for individual use and "Ask Your Data" custom-GPTs for public or organizational use. It also provides the option to search the web for information, which can be useful not for private data, but for up-to-date information released after the model's training knowledge cutoff. Interestingly, these web searches can be initiated explicitly or, depending on the prompt, automatically by the system, but a later chapter will discuss how these automatic RAG triggers work within the framework of AI agents. While all these built-in methods are highly user-friendly, they offer limited control over retrieval quality, and both the database and model are retained by the chat provider. In the worst-case scenario, private inputs could be used in future model training, so it is essential to carefully review the terms of service. Alternatively, for enhanced data governance, major cloud providers like Microsoft Azure, Amazon Web Services (AWS), and Google Cloud now offer no- and low-code RAG solutions, integrating managed vector stores and LLMs within an organization's dedicated resources. Finally, a completely custom RAG solution can be developed through programming. Open-source frameworks like LlamaIndex and LangChain can make this task easier by providing predefined functions for retrieval algorithms and LLM interactions. Vector embeddings can be stored on-premises or with a trusted third-party provider. LLM access can be obtained through hosted API services, like OpenAI's ChatGPT or Anthropic's Claude, or through locally deployed models. Although this approach requires significant development effort, it offers the highest level of flexibility, customization, and control over data privacy.

Fine-tuning, previously introduced in the context of training LLMs, offers another way to enhance the context of pre-trained LLMs. While RAG adds external context during query execution, fine-tuning integrates specialized information directly into the model. Figure 9 shows a fine-tuning process for a chatbot designed to aid in creating LEED certification review reports in order to obtain a green building certification. In this framework, the model takes key project data as input and generates a draft review that aligns with LEED standards. In order to achieve this, initially, a collection of existing review reports is compiled and converted into prompt-response pairs (typically several hundreds of them). These examples are used to further train (fine-tune) a pre-trained LLM, ensuring that LEED criteria and report format are embedded directly into its parameters. Since the model contains the necessary domain knowledge, there is no need for external information retrieval during inference—though adding a system prompt summarizing the LEED guidelines can help prevent hallucinations. After fine-tuning, the user simply provides project details, and the chatbot produces a structured LEED review report.

Fine-tuning an LLM is typically more intricate than setting up a RAG system, as it involves tasks like data cleaning, tokenization, and example formatting, which often require scripting. Nevertheless, there are no-code graphical user interface services available for fine-tuning, such as OpenAI's Fine-tuning Playground, Azure AI Studio, or Hugging Face AutoTrain, that simplify dataset uploads and training monitoring. However, most production teams opt for API- or library-driven pipelines to have better control over the delicate fine-tuning process. Once training is completed, the custom model can be deployed similarly to any other LLM or RAG endpoint—whether on a third-party platform like OpenAI's,

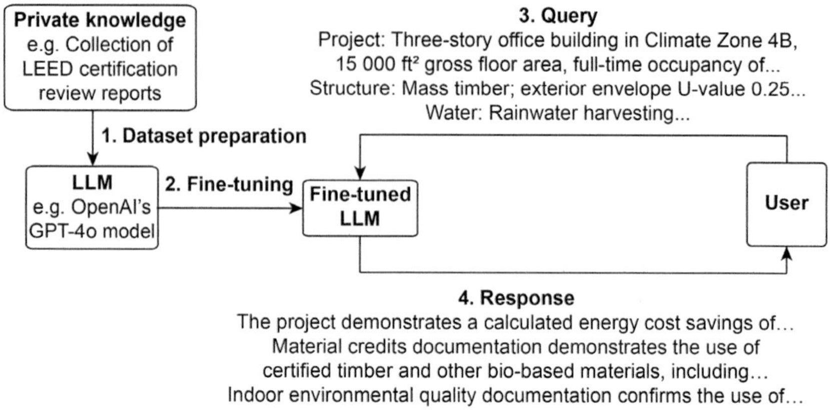

Figure 9. Fine-tuned LLM for automated generation of LEED certification review reports.

as a managed service from a cloud provider for a more enterprise-focused solution, or locally if using an open-source model. Finally, it is also possible to pre-train an LLM from scratch with custom training data for organizations seeking maximum control over the inference of their LLMs. GPU-rental platforms such as Lambda Labs provide on-demand access to multi-GPU servers—avoiding the capital expense of owning a cluster while still permitting fully private model training.

In summary, choosing between RAG and fine-tuning to integrate sustainability expertise into a chatbot largely depends on the specific use case. RAG provides a straightforward and dependable method to access a constantly updated private database. For instance, the previously mentioned RAG bot used for retrofitting and renovation projects can easily incorporate new projects, thereby enhancing its understanding of structural solutions for existing buildings. However, RAG has the downside of increased latency due to the additional retrieval step, which can slow down response times. Alternatively, organizations willing to commit more resources, both financially and in terms of time to prepare the training data, might opt to fine-tune an LLM. This approach not only embeds specialized knowledge into the model but also trains it to output information in a preferred style and format, such as generating LEED certification review reports. In this context, the reference database does not require frequent updates, as new reports would not provide much additional insight into the LEED guidelines. Furthermore, fine-tuning leads to much faster inference than a RAG pipeline since the model can respond directly without needing an extra data retrieval phase.

Code Vibing Data Exchange for Life-Cycle Assessment Dashboards

LLMs are not just good at writing prose or answering questions. They have seen so much code during training that they can actually write programs too. Although complex applications with codebases of tens of thousands of lines remain a challenge for LLMs because of context-window limits, the "vibe-coding"[11] approach is very successful for early prototyping. Users simply tell the model in plain words what they want it to do, and the AI turns their description into working code. There are also built-in features in modern chatbot interfaces that support and enhance the "vibe-coding" approach: Anthropic's Claude lets users generate and manage self-contained code snippets called Artifacts in a side panel—so they can tweak them, test them, and plug them straight into their workflows without ever leaving the chat and OpenAI's Canvas provides an interactive whiteboard where users can sketch user interface (UI) layouts or flow diagrams, have the model translate them into code, and iterate visually in real time.

Throughout the building design process, there are numerous chances where simple code snippets can be highly beneficial. One of the most common uses for these snippets is facilitating data exchanges between different software systems. For example, exchanging BIM models created with proprietary BIM software among various companies. Another example is creating an interface between BIM and FEM calculation software, as discussed in the interview with Alexander Stirken in the previous chapter Additionally, in terms of sustainability, these snippets can help import 3D model data into business intelligence dashboards for LCA visualizations. These dashboards, like the one depicted in Figure 10, are extremely useful to keep track of the sustainability metrics of the project throughout the design phase and can be shared among different project stakeholders (Apellániz et al. 2024) to facilitate the collective decision-making process. Just as McKinsey's Catalyst Zero (McKinsey and Company 2025) automates the cleaning and structuring of Enterprise Resource Planning (ERP) data for marginal-abatement cost curves, an LLM can be prompted ("vibed") to ingest raw LCA outputs coming from a certain LCA software, normalize and aggregate the fields needed, and emit ready-to-load tables for the dashboard.

[11] The term "vibe coding" first appeared in early 2025, coined by AI researcher Andrej Karpathy to describe a shift from writing every line of code by hand to simply "vibing" with a model via natural-language prompts (Roose 2025). Drawing on advances in large language models like OpenAI's Codex and ChatGPT, Karpathy envisioned developers conversing with AI – much like chatting with a teammate – to generate, refine, and test code on the fly. This conversational approach laid the groundwork for today's accessible, prompt-driven programming workflows.

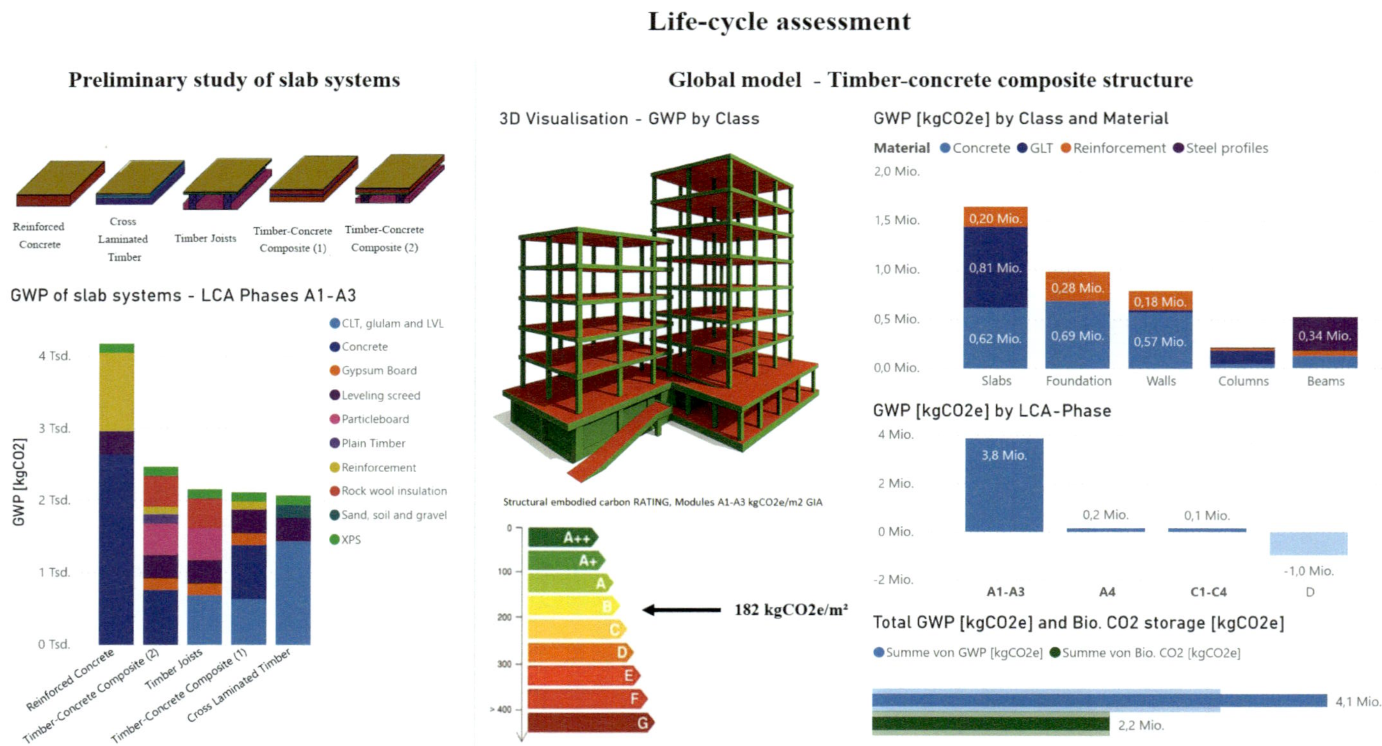

Figure 10. LCA dashboard with Power BI.

Multimodal Generative AI

6

Generative-image tools emerged just months before ChatGPT, collectively sparking what many now refer to as the Generative AI Revolution. In April 2022, OpenAI introduced DALL-E 2, followed by Stability AI's open-source Stable Diffusion in August; and suddenly, anyone could create photorealistic images with a simple text prompt.

Initially, hobbyists flooded the internet with sketches and memes,[1] but soon enough, professionals such as marketing designers and artists[2] experienced the impact of AI-generated images reshaping their work. The architecture sector was not immune to this change either. Architectural firms started using AI to create photorealistic visualizations that helped them promote their ideas and win commissions. Traditionally, building visualizations involved rendering[3] a 3D model, providing a direct link between design and visualization. However, with AI, this connection is not always necessary. Therefore, these quick visuals allowed designers to rapidly iterate through multiple design possibilities, but they also opened the door for presenting deceiving proposals, such as visually appealing yet structurally impossible designs, or worse, projects that only seem sustainable.

[1] These early AI-generated images blended amateur creativity with absurdist humor, producing viral hits like the "Valenciaga Pope," a satirical portrait of Pope Francis clad in high-fashion couture riffing on Valentino and Balenciaga.

[2] In August 2022, Jason M. Allen's AI-generated image "Théâtre d'Opéra Spatial" won first prize in the Colorado State Fair's digital arts category, igniting fierce debate over AI authorship and the fairness of juried art competitions (Vincent 2022).

[3] A render is a computer-generated image produced by a graphics engine that simulates light, materials, and perspective to depict a 3D model realistically.

DOI: 10.1201/9781003506324-6

This chapter begins by examining the impact of diffusion models, a deep learning technology, on AI image generation. It continues with a discussion featuring Cas Esbach, an architect and prominent expert in AI visualizations, who explains how to integrate these visuals into the design process to support honest storytelling in sustainable projects. The chapter then explores additional workflows in multimodal generative AI, such as image-to-text methods that connect the built environment to LLMs, along with several other emerging approaches.

Text-to-Image: Enhancing Sustainable Building Storytelling with Diffusion Models

Building visualizations serve multiple purposes: they are essential tools in design competitions, client presentations, and the early phases of design. Developers use them to assess whether a project will appeal to future buyers or tenants, while planning authorities evaluate them to determine how well the proposal integrates into the urban environment and complies with regulations. In these contexts, architects create visualizations to strengthen their proposals and secure commissions. Once selected, they continue to refine and present these visuals, enabling clients to explore different design alternatives. Traditionally, such visualizations were produced by rendering a 3D model manually constructed in a CAD program. This process is computationally intensive, as the software must calculate how each ray of light interacts with the scene's polygons and textures through bouncing, refraction, and shading. Consequently, the workload is typically offloaded to GPUs, whose parallel cores calculate the final color for thousands of pixels simultaneously. However, in 2022, the introduction of text-to-image generative AI tools in public beta significantly altered the workflow for architects. They could now produce nearly photorealistic concept visuals simply by describing a building in natural language. While this development did not render traditional 3D workflows obsolete, it made concept design visualization much faster and more accessible.

As with LLMs, two distinct approaches have emerged: proprietary cloud services, such as OpenAI's DALL·E 2 (released in closed preview in April 2022) and Midjourney (open beta in July 2022), and locally runnable open-source models like Stable Diffusion (public release in August 2022). However, in contrast to LLMs, where most users prefer proprietary cloud services such as ChatGPT or Google Gemini, open-source image-generation models have become particularly popular in architectural practice. Because their weights and code are publicly available, firms can fine-tune these models using proprietary project data while keeping everything on-site. This setup allows for greater creative control and improved data privacy. Since running these large models locally requires high-end GPUs, many firms have repurposed existing workstations originally purchased for rendering instead of investing in new hardware.

While this new, rapid, and widely accessible workflow for image creation enables architects to showcase sustainable projects more effectively, potentially leading to more commissions and the promotion of sustainable architecture (assuming most architects advocate for sustainability), it demands careful implementation. Figure 1 depicts a building visualization produced by Stable Diffusion v1-4 using the prompt: "modern glass skyscraper, two offset rectangular volumes stacked with a clear open gap in the middle, dramatic cantilever, high detail architectural rendering, photorealistic, 8k resolution, golden hour sunlight, warm reflections on curtain wall, cinematic city skyline in background, subtle aerial haze, physically accurate shadows." The image was generated in mere seconds without the need for an underlying 3D model, significantly reducing the time compared to traditional rendering methods. However, despite its impressive appearance, the result, among other issues, raises structural concerns: the vertical load path is interrupted at the gap, the dramatic cantilever lacks evident support, and the design offers no apparent system for lateral stability. Even though the concept is not feasible for construction, an unethical architect might attempt to market the void as a "sustainable" feature enhancing natural ventilation. Regarding this dilemma, the next section's interview with Cas Esbach will demonstrate how to use AI-generated images honestly to propose and illustrate genuine sustainable projects. But first, it is essential to examine how these AI models function to grasp their capabilities and limitations.

Most state-of-the-art text-to-image AI tools are based on diffusion models. The core architecture of many of these deep learning models is a U-shaped convolutional neural network (U-Net) (Ho et al. 2020), which uses convolutional layers to capture local image features, similar to the CNNs introduced in Chapter 4. In diffusion models, the U-Net backbone is enhanced in two ways: first, by applying self-attention at intermediate spatial resolutions to model dependencies between distant pixels, in a manner similar to how LLMs use attention across tokens to capture long-range context; and second, by incorporating cross-attention layers that condition the diffusion process on a text prompt, which is fed into the network as an embedding vector (Rombach et al. 2022). For example, Stable Diffusion generates this embedding vector using the CLIP model (described in Chapter 5), which was trained to produce embeddings for both images and text simultaneously.

Figure 2 illustrates the process by which the Stable Diffusion tool generates an image of a family house, demonstrating how diffusion models function. Starting from random noise; which is analogous to an image where every pixel has arbitrary RGB values; diffusion models gradually

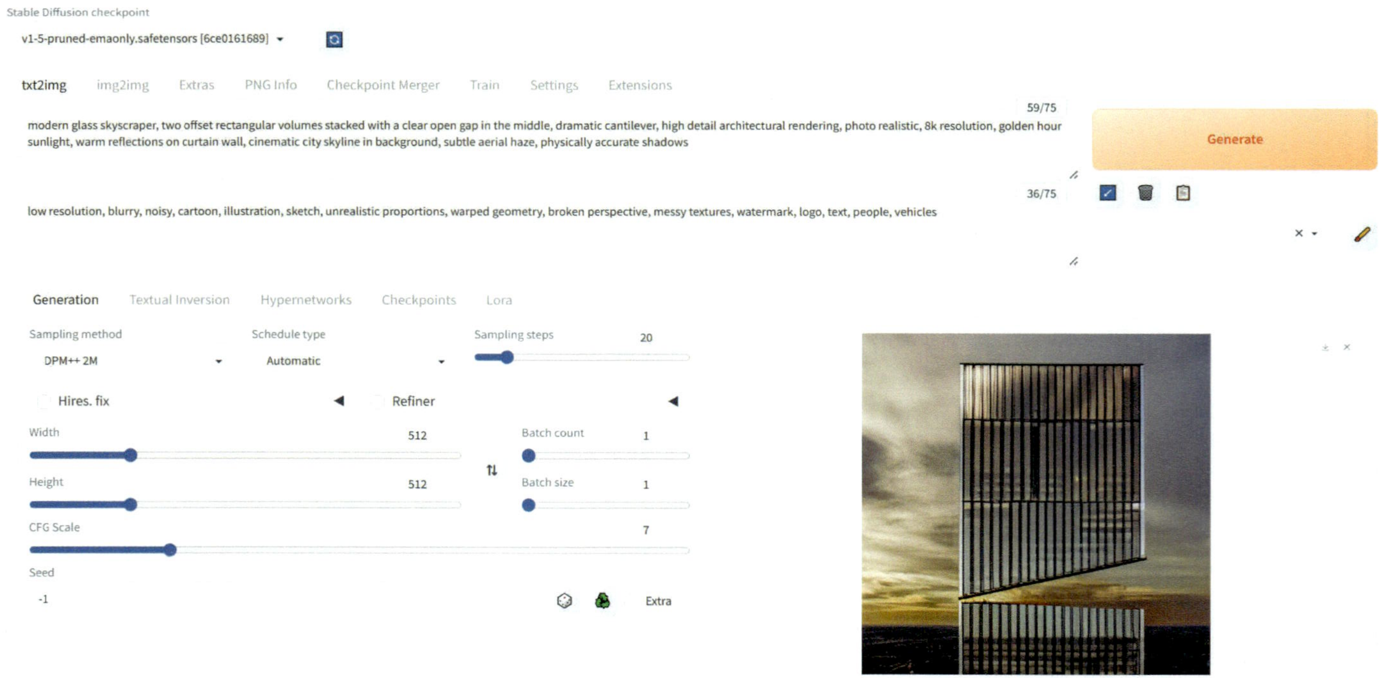

Figure 1. AI-generated skyscraper visualization with AUTOMATIC1111's Web UI.

remove this noise in incremental steps. This denoising process is guided by a text prompt and continues until the final image emerges. The training process explains how the model learns to perform this transformation. Firstly, the training data for diffusion models is generated by taking clean images and gradually corrupting them by adding increasing amounts of random noise step by step. This is known as the "forward diffusion" process, which typically occurs over 1,000 steps, indexed from t = 0 to t = 999. The model is then trained to perform the reverse process, known as "reverse diffusion". Given a noisy image and the corresponding step t, the model learns to predict the noise that was added so that it can be removed, thereby recovering a previous state of the image. This strategy of predicting noise, rather than the clean image itself, has been found to yield better results in practice. Additionally, the algorithm that determines the number of denoising steps and the size of each step is called the schedule, and it plays an important role in the model's performance. Finally, to guide the process using prompts, referred to as "conditional diffusion", the training data pairs each image with one or more text descriptions. This enables the model to learn not only how to denoise but also how to do so in a way that aligns with the given prompt.

Given this explanation, why in Figure 2 the starting frame at $t = 999$ does not resemble random colored pixels, but instead appears as blurry, blueish noise? The reason is that some tools post-process the image generated by the diffusion model. For example, the family house shown in Figure 2 was created using the Stable Diffusion tool that performs computations not on full-resolution images but in a compact latent space with dimensions 4 × 64 × 64. This latent output is then passed through a second neural network, known as a Variational Autoencoder (VAE), which transforms it into a full-sized image with dimensions 512 × 512 pixels and three color channels, forming a 3 × 512 × 512 representation. Operating in this smaller latent space is nearly 50 times more efficient than processing full-resolution images directly, and it also improves the

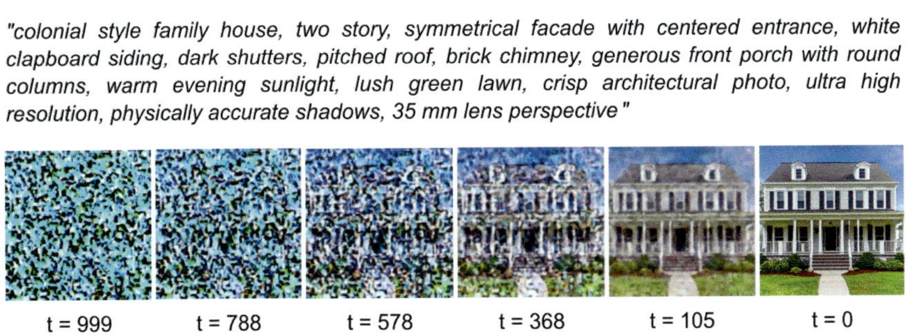

Figure 2. Image generation process with diffusion models, generated with Stable Diffusion v1.5 (Stability AI).

quality of image generation. For these reasons, the model is designed to work in latent space. As a result, the frame at t = 999 in Figure 2 does not resemble random colored pixels, because what is shown at that stage is decoded latent noise rather than raw RGB noise.

With the explanation of the functioning and training of diffusion models completed, it is possible to assess their strengths and weaknesses. On the one hand, the U-Net architecture, which combines convolutional layers for capturing local image details with self-attention mechanisms for modeling relationships between distant pixels, is crucial for producing images that are so realistic and detailed that they could be mistaken for photographs. These models can mimic a wide array of styles and subjects, ranging from classical art to futuristic scenes, as long as similar examples are present in their training data. As shown in Figure 2, they can generate precise design visuals when given clear prompts, but only if the desired output does not significantly deviate from what the model has previously encountered during training. On the other hand, their creative capabilities are largely dependent on the patterns within the training data. While they learn to predict the noise added to original images, they may struggle with highly regular or intricate structures, like steel truss systems, or unique structural designs that are not well represented in the training set. Representing such elements often requires a level of reasoning or abstract composition that simple denoising based on pixel statistics cannot fully achieve, which also accounts for their difficulty in accurately rendering text. Additionally, very fine details specified in prompts may be lost. Converting text into a fixed embedding vector can reduce the precision of highly specific instructions, so an instruction like "a building with exactly seven small, square windows on the north facade" might be challenging. The cross attention mechanism tends to highlight the most prominent ideas from the prompt rather than every numerical or relational detail, and the generation process occurs in a compact latent space (for example, Stable Diffusion's 4 × 64 × 64 representation), which, although efficient, can further reduce fidelity when encoding and decoding images with highly specific details.

Customizing diffusion models through fine-tuning is becoming a vital tool for architecture firms working with AI-driven visualizations. While many public models are available, including those trained mostly on architectural photography, they often fail to capture a specific firm's material palette, spatial logic, or design philosophy. For instance, when architects at MVRDV used the prompt "an MVRDV style building" in Midjourney, the result was a brightly colored, voxel like structure that misrepresented the studio's design approach (Show it better 2024). Fortunately, diffusion models can be fine-tuned to incorporate a firm's unique design language, creating images that accurately reflect its architectural style. Techniques like Low-Rank Adaptation (LoRA) and similar adapter modules add a few trainable parameters to a fixed base model. Instead of updating millions of weights, LoRA breaks down the weight updates into low rank matrices, significantly reducing computing, memory, and data requirements. By training just these additional adapters with a curated set of firm specific images, it is possible to produce a lightweight fine-tuned model that embeds the studio's stylistic essence into every generated visualization without retraining the entire network. Prompt tuning offers a complementary method to adapters: it customizes the model by enhancing the input prompt's embedding vector with an additional vector learned during fine-tuning, thereby keeping the model's parameters unchanged. In order to experiment with fine-tuning of diffusion models, the Stable Diffusion open source models are a remarkable choice for applying adapter modules and prompt tuning, as they provide complete access to their weights and training scripts and are computationally efficient, achieving high quality results in a compact latent space. Additionally, the active ecosystem of tools and services developed around Stable Diffusion; such as AUTOMATIC1111's Web UI, ComfyUI, Hugging Face's inference API, Stability AI's DreamStudio, and various cloud hosted platforms; simplifies both training and inference.

Before concluding the text-to-image workflow, it must be mentioned that diffusion models are not the only AI approach for image generation. An earlier approach was GANs, which, like diffusion models, start from random latent noise but use a generator network to produce an image and a discriminator to judge its realism. Training these two in opposition can result in a powerful model but is often unstable, while diffusion models tend to converge more reliably. Another AI approach for the text-to-image workflow is autoregressive image models, first popularized around 2016. They generate images by predicting one pixel or a patch of pixels at a time based on the previously generated ones, whereas diffusion models denoise an entire image in parallel over hundreds of steps. OpenAI's March 2025 GPT 4o image generator adopts this autoregressive approach natively in the chat interface (OpenAI 2025a), though its full architectural specifics remain undisclosed. To sum up, innovation moves at breakneck pace, and the text-to-image workflow continues to improve in fidelity, speed, and user control.

Image-to-image and 3D-to-image: Towards Honest Architectural Visualizations

The purpose of architecture is to transform abstract concepts into a physical building. When creating visualizations of buildings designed for sustainability, such as those using biogenic materials or repurposing existing structures, it is essential that the design is feasible. The layout and overall dimensions, along with the efficient arrangement of spaces for its occupants, must be accurately represented in the visualization to provide a truthful impression of the potential building. While text-to-image methods might suffice for exploring initial concept ideas, achieving detailed architectural visualizations that offer an honest depiction requires more than just text prompts and high quality training data. Fortunately, there are various techniques that provide AI visualizations with the necessary level of precision in the context of image-to-image and 3D-to-image workflows.

One method to moderately influence the outcome of AI visualizations is through Image Prompt adapters (IP-adapters) (Ye et al. 2023). These adapters allow the style or composition of a reference image to be replicated. They function as a Stable Diffusion add-on, enabling the use of images as prompts, similar to proprietary tools like Midjourney and DALL·E 3 do, though those platforms do not disclose their techniques. The process involves inputting an image, such as a sketch or another building visualization for style reference, through a pre-trained image encoder, for example the CLIP model. The IP-adapter model is then trained to transform the resulting feature vectors into embeddings that are more effective for directing the image generation process. The idea is that it is more efficient to train an adapter than to retrain the entire image encoding pipeline. These embeddings are then appended to the embedded prompt text to guide the Stable Diffusion model during image creation more precisely than a text prompt alone. This method is useful for style transfer or for establishing a consistent material palette or specific visual look across several AI generated images. For example, a picture of a previous building designed in the office could be used to ensure that all AI generated visualizations maintain a consistent style. This technique is also valuable for producing reference guided variations, such as different seasons, lighting conditions, or materials of the original image or visualization.

Another method offering more detailed control than IP-adapters is ControlNet (Zhang et al. 2023). This is an extension of diffusion models that enhances the standard conditioning on text embeddings by incorporating a parallel copy of the U-Net with trainable convolution layers. These copied layers are trained while the base U-Net remains unchanged, enabling the model to learn precise spatial conditioning (such as edge maps, semantic layouts where different colors signify various materials, surface normals, and basic line sketches) without losing its original ability to generate architectural visualizations. During each denoising step, the model considers both the usual text embeddings and the ControlNet features, ensuring that the final output aligns with the user supplied structure or guide. While various open-source tools like the AUTOMATIC1111 web UI and ComfyUI offer ControlNet modules,

Figure 3. Sketch to render workflow with the Vizcom.ai tool using ControlNet-style conditioning under the hood.

there are also proprietary tools that utilize the ControlNet method. Figure 3 illustrates the creation of an AI visualization using the Vizcom.ai tool, guided by both an input text prompt and a handmade sketch, so can be noticed how the structural elements of the timber canopy maintain their dimensions and arrangement as depicted in the original sketch. Another relevant case study would involve exporting an edge map or a semantic image as screenshot from a 3D model to generate a realistic visualization consistent with the design, bypassing the need to manually assign materials to each object in the model as traditional rendering requires. Furthermore, combining the ControlNet and IP-adapter methods is possible, with the former offering control over the structure and layout of the visualization and the latter over the style. This technique can be used, for example, to create multiple consistent views of the same building, which is essential in architectural competitions.

Finally, conventional renderings can perfectly coexist with these AI approaches. While they have the drawback that the user still needs to assign materials and wait for render times, the result includes one hundred percent physically accurate shadows, lighting, and reflections, which is especially useful when the user wants multiple camera angles or precise sun studies. For instance, in the previous image, it can be noticed that although the structural elements are accurately represented, the shadows cast by the different canopy columns are not entirely consistent. Regarding the so-called AI rendering tools, like "Veras AI," which work as plug-ins integrated into 3D modeling software, although most of them are proprietary and do not reveal how they work under the hood, they likely use a ControlNet approach to produce renders: for example, by exporting a depth map from a certain view of the 3D model to condition the generation of the AI visualization based on the user's prompt.

Last but not least, although AI approaches that grant users control over generated images are evolving at an impressive pace, there is still a need for post processing in 2D drawing software (for example, Photoshop) to remove minor glitches, insert realistic human figures, apply lighting or color filters, and so on, especially if a high-quality, professional-level submission is required.

Interview with Cas Esbach on how state-of-the-art AI Visualizations can Catalyze Sustainable Building Projects

DA: At the time of this interview, you are a project leader working in Rotterdam with a leading architectural practice and have also become one of the leading voices in the relatively new field of architectural visualizations using AI. How and when did you get into AI visualizations? Are you a tech enthusiast like some of your colleagues in experimental design and research groups, or do you primarily appreciate the easy access to AI visualization tools, meaning there is no need for coding or extensive additional software knowledge?

CE: I started working with generative models in 2022, and from the start, I was captivated by the way they could enhance the design process. I would not say I am as much of a tech enthusiast as some of my colleagues in R+D, but I do appreciate how technology can empower creativity. What drew me into AI was the ease of use, the immediacy, and how it helped us communicate ideas more effectively. It was thrilling to be able to take an abstract concept and quickly turn it into something visual without the lengthy rendering times of traditional methods. AI made it possible to explore many variations almost instantaneously, and that kind of creative freedom is hard to resist. So, while I'm not someone who is deeply into coding or the backend tech, I do have a strong appreciation for the creative and practical potential that AI brings to architecture.

DA: Your office in Rotterdam has certainly built a big reputation for applying state-of-the-art technology in their designs. The Market Hall in Rotterdam, for instance, features the largest mural in the Netherlands on the inside of the building: an 11,000 m^2 artwork printed by Pixar from a 1.47-terabyte file (back in 2014!). Today, if I remember correctly, you have in your office a room full of GPU-powered computers exclusively dedicated to architectural visualizations. Are you still using rendering engines, or do you exclusively produce AI visualizations now? How much time do you spend in this room?

CE: We have a setup of older rendering machines that everyone in the office can access. I do not have the exact specs in mind, but they are quite powerful and handle our visualization needs well. These machines were originally brought in for traditional rendering, but we've now repurposed them to work with AI-based workflows. It is a practical evolution of our resources: we already had GPU-powered systems, so we adapted them for Stable Diffusion and other AI models, which made sense both economically and technically. We still use traditional rendering engines, especially for projects that require a high level of detail and realism. It is not about replacing traditional methods entirely; it's about expanding our toolkit so that we can choose the most effective approach for each stage of the project. The aim is to balance efficiency, creativity, and the technical needs of each project.

DA: Speaking of balancing the best tool for each purpose, what is your favorite tool for the early sketch phase? In this context, do you view AI as just a tool or as a co-designer?

CE: For early sketching and any AI image generating, Stable Diffusion is my go-to tool. It offers an unparalleled level of freedom and adaptability, which is crucial in those initial phases of a project. What is powerful about Stable Diffusion is that it is open source and constantly evolving, making it easier to align the generated visuals with our vision and design language. Right now, within the practice, AI is more of a tool: something that helps us articulate ideas, test variations, and communicate more effectively. But I genuinely hope that in the future, AI can evolve into a true co-designer. Similar to how I use ChatGPT as a co-writer for my emails, I would love to see AI in architecture go beyond just generating options to making informed suggestions based on our past projects and lessons learned. That kind of partnership, where AI becomes an active collaborator in the creative process, is something I find incredibly exciting.

DA: After the initial conceptual sketch phase, you need to work on project visualizations that convey a feasible design to the client. Which tool or workflow do you use at this stage? Do you make sure that those visualizations are based on e.g. a realistic building structure so they can end up being built?

CE: Once we move past the initial conceptual phase, the workflow becomes a bit more hybrid in the design teams. We primarily use Stable Diffusion in combination with other techniques to make sure the visualizations meet the project's requirements. Often, this means blending AI-generated outputs with traditional 3D rendering and further refinement in Photoshop. The aim is to create visualizations that are evocative and convey the atmosphere of the design while staying grounded in what is technically achievable. AI helps set the mood and helps us iterate quickly, but we always ensure that the visuals we present to clients are rooted in buildable reality. It's a balance between leveraging AI for its creative power and maintaining the precision needed to ensure the project can move forward seamlessly into construction.

DA: You mentioned before the importance of "aligning the generated visuals with our vision and design language". Can you elaborate on this?

CE: Training our own models gives us a level of control and consistency that is not possible with off-the-shelf options. Public models can be great, but they often do not align perfectly with our specific aesthetic goals or the design flexibility we need. By training our own models, we can fine-tune them to fit our projects, ensuring a consistent design language across different visualizations. It is about maintaining the firm's ethos while exploring new possibilities. Sharing some of these models publicly is a way for us to connect with the broader design community. There is a balance we are trying to strike between being open and contributing to collective knowledge while also protecting our unique methods and intellectual property. Getting feedback and seeing how others use our models helps us push the boundaries even further, which is always beneficial.

DA: Unfortunately, not all architectural firms share such an ethical approach towards the use of AI.

CE: Absolutely, AI can be used for greenwashing. AI-generated visuals can be incredibly compelling, and it is easy to make a project look more sustainable than it actually is. It is a bit like the issue of misinformation on social media: the visuals might be beautiful, but they can sometimes create unrealistic expectations. Transparency is crucial here. Firms need to be clear about what is feasible and what is purely conceptual. Ethical guidelines for using AI in architecture are essential to make sure we are not misleading stakeholders. AI should help us achieve sustainable goals, not just create the illusion of them. In the end, AI is just another tool: like a sketch, a render, or a piece of writing. It is all about how you use it and maintaining integrity in the design process.

DA: We need to address the elephant in the room before finishing the interview. The use of AI, in general, and AI visualizations, in particular, consume a considerable amount of energy and thus generate CO_2 emissions.

CE: I am very aware of the energy demands and carbon footprint associated with AI, and it is something we take seriously. The models we use, especially as they grow more complex, require significant computational power, which does come with an environmental cost. At the same time, these tools bring immense value: they allow us to iterate faster, explore different scenarios, and make more informed design decisions. This kind of rapid iteration is particularly useful when it comes to sustainable building designs. By being able to test various spatial configurations and material choices quickly, we can better optimize for sustainability right from the start of the project. It is all about striking the right balance. As this technology continues to evolve, we are seeing improvements in energy efficiency, and we are mindful about using AI where it truly adds value. I think the industry as a whole will continue to find ways to mitigate the environmental impact as these tools become more widespread.

Image-to-Text: Bridging the Gap Between the Built Environment and Large Language Models

The interaction with the built environment holds great potential for improving sustainability and reducing the building industry's carbon footprint. Previous sections have already emphasized the importance of building renovations, extending a building's life cycle, quality control, and facility management. However, despite all the advances in generative AI, applying this evolving technology to the built environment remains a significant challenge. AI systems typically expect text inputs that can be tokenized and processed by transformer architectures, whereas we understand and describe most existing buildings through drawings and images – especially those constructed before BIM models became common (even today, BIM is not always used by default).

Fortunately, vision-language models (VLMs), which combine computer vision and natural language processing, may be applied to this challenge. OpenAI's GPT-4 (launched March 14, 2023) was the first GPT model designed to accept both text and image inputs. Since then, different proprietary services and open-source models have appeared that implement this technology. Essentially, a VLM is built on a standard LLM but adds a vision encoder to handle image inputs (see model architecture in Figure 4). The vision encoder, which is often a CNN or Vision Transformer (ViT), has been trained on millions of image-text pairs, giving it the ability to associate images and text. Because those raw feature vectors do not match the LLM's token format, they're then fed through a small adapter (much like the IP-adapters described in the "image-to-image" section) that converts them into "image tokens" in the same embedding space the LLM uses for words. Finally, the LLM processes those image tokens together with any text tokens in its input. The combined sequence of tokens passes through the LLM's transformer layers, and it generates a text response just as it would on a pure text prompt. Only now, it is conditioned both on the image content and on any accompanying text.

Focusing on the computer vision part, input images are typically scaled down or center-cropped first, for example, OpenAI's vision models resize an image so its shortest side is at most 768 pixels, whereas a modern smartphone photo can be 4032 × 3024 pixels. A ViT then divides the image

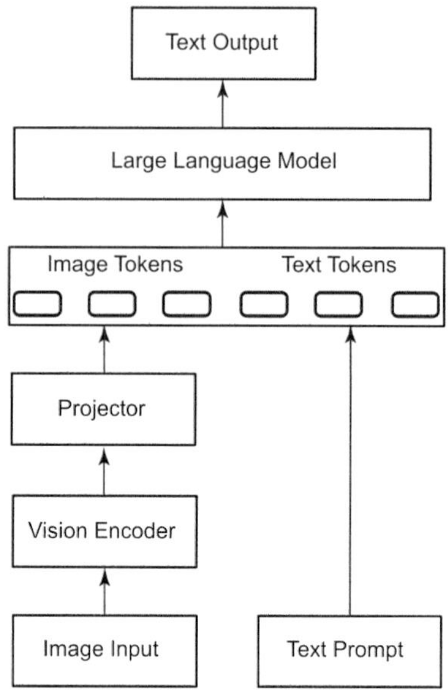

Figure 4. Common architecture of vision language models.

into a grid of fixed-size patches (commonly 16 × 16 pixels, sometimes 14 × 14). Each patch is flattened and converted into a numeric vector derived from its RGB content. Positional embeddings are added to each patch vector, so the network knows where it belongs, and an extra vector derived not from the patches but from the whole image, called the class vector, is prepended to the sequence to capture a summary of the entire image. This full sequence is then passed through a stack of self-attention layers (many VLMs reuse a frozen CLIP ViT here, so the vision backbone itself need not be retrained) to generate contextualized embeddings that capture the image's semantic content. Because self-attention allows every patch attend to every other patch, the model can capture long-range patterns, such as the vertical alignment of windows or the contrast between roof and façade, without relying on the rigid hierarchical filters of older CNNs. Finally, the resulting image embeddings are routed through a small adapter and fed into the LLM, which processes them alongside any text tokens to generate the response.

In general terms, VLMs can be used for visual questions-answering, image summarization and parsing text and handwritten documents. Figure 5 evaluates the response of the VLM o4-mini-high when asked to identify and describe a building pathology related to the presented column (in this case studies, responses generated with the more advanced GPT-5 model did not show a significant difference compared to those from o4-mini-high). The picture could be tricky to an AI model due to the paint coating that conceals the structural concrete. Nevertheless, the

AI computer vision evaluation	
Vision language model: OpenAI's o4-mini-high (ideal for visual reasoning)	
Feature	**Evaluation**
Description of the building pathology	✓✓✓
Cause of the building pathology	✓✓
Required action	✓

Figure 5. Evaluation of a vision language model on a building pathology example.

model correctly identifies the column as a structural reinforced concrete column and successfully identifies the concrete spalling pathology. However, it is not certain about the cause of the pathology, although it guesses a poor concrete mix with too much water. Despite much more exhaustive testing would be required, it could be argued from the response of the recent o4-mini-high to this challenging picture that VLMs have a potential to be applied for the identification of structural pathologies. Human supervision should be provided for analyzing the cause of the pathology and determining the appropriate corrective action, although system prompting or even fine-tuning would likely improve the model results. Additionally, the identification of building pathologies does not need to happen in milliseconds, since responding to them and carrying out retrofitting may take from hours to weeks. Therefore, it should be acceptable that reasoning models like o4-mini-high require several seconds to produce an answer, unlike in other computer vision domains, such as self-driving vehicles, where AI responses are expected in milliseconds.

Figure 6 shows the evaluation of the response of the same VLM, o4-mini-high, when it is presented with a moderately small floor plan. The model successfully identifies all the room labels, even though they are handwritten, and can understand the layout to a good extent: where the rooms are, how big they are, what the overall spatial logic is, etc. Optical Character Recognition (OCR) is the process of converting images of text into machine readable characters. VLMs do not interpret text labels character by character, but rather from the broader context of the image, as illustrated previously in Figure 4 on the model architecture.

This approach can be effective for understanding whether a room is a bedroom or living room based on visual labels, but it may lead to some information loss compared to traditional OCR tools when interpreting isolated annotations or dimension figures in technical drawings. Furthermore, in this case study, likely due to the resolution reduction

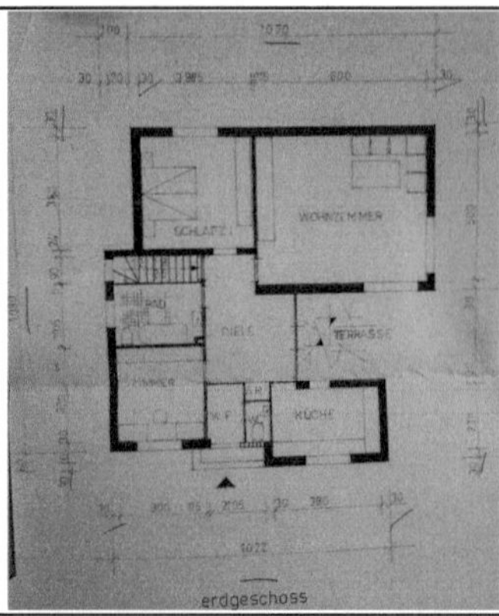

AI computer vision evaluation	
Vision language model: OpenAI's o4-mini-high (ideal for visual reasoning)	
Feature	**Evaluation**
Identification of the room labels	✓✓✓
Interpretation of the floor layout	✓✓
Point coordinates for digitalization	×

Figure 6. Evaluation of a vision language model on a hand drafted floor layout.

performed by the VLM tools on the input images, the model was not able in this case study to reliably recognize the dimension lines or determine which building elements each dimension refers to. This makes the task of digitizing hand drafted floor plans into a usable digital format still very challenging for VLMs—despite its potential to simplify renovation projects significantly.

While plan interpretation and error checking would still require human supervision, processing technical drawings for simple tasks, such as carrying out as built LCAs where the main goal is identifying building materials, is certainly feasible using VLMs. Finally, the model's response also showed lack of advanced spatial reasoning (such as optimal window placement or furniture arrangement), which is also recognized by Nvidia as one of the current limitations of VLMs (NVIDIA n.d.). For example, the startup "syte", which claims to use AI to evaluate the potential of parcels and existing buildings and won the German AI Startup Award in 2023, allegedly relies on LiDAR point clouds and land registry data instead of using AI computer vision to interpret the geometry of the built environment. Nevertheless, VLMs represent one of the fastest evolving areas in AI, with new approaches constantly being researched by major AI companies (Shi et al. 2024), and significant progress in this domain is expected in the near future.

Other Multimodal Workflows

In addition to the approaches discussed above, several other multimodal generative AI workflows, though not yet directly applied to building sustainability, are already mature in other domains and hold significant mid-term promise for the construction sector.

A multimodal workflow with major potential is audio-to-text conversion using automatic speech recognition (ASR). The average speaking rate for English speakers is about 150 words per minute, compared to roughly 40 words per minute for typing, making speech-to-text a far more efficient input modality. When applied to everyday computing tasks, such as composing emails, drafting reports or logging meeting notes, this approach can significantly accelerate project planning and coordination, freeing up more time for sustainability analysis. Moreover, pairing voice input with AI agents could eventually enable fully vocal control of software interfaces, as the next chapter will explore. Modern transformer-based automatic speech recognition (ASR) models, such as OpenAI's Whisper, now transcribe speech in over ninety languages with high accuracy and robustness to accents and background noise.

Another rapidly evolving workflow is text-to-3D generation. The foundational technique of Score Distillation Sampling (SDS) creates an initial 3D model by generating multiple camera views with a diffusion model and then refines that model by comparing rendered views with the diffusion outputs. Follow-on methods like have appeared since then and several commercial platforms (e.g., Meshy.ai, Kaedim) now offer high-quality "text-to-3D" services, though they keep their internal architectures proprietary. However, while these pipelines can produce impressively textured and stylized 3D models in minutes, their geometric accuracy (in terms of exact dimensions, topology consistency, etc.) remains too coarse for using such an AI-generated building 3D model for a rigorous LCA or structural analysis. Nevertheless, they can be used to enhance clarity of the results, for instance by complementing a predicting a carbon footprint in very early stages (see Chapter 4) so that the reader can more easily understand the kind of building whose carbon footprint is being forecast.

Last but not least, the text-to-video workflow is rapidly evolving. OpenAI's Sora, released publicly in December 2024, can produce up to one-minute clips directly from text prompts. However, despite the remarkable resulting videos, they often exhibit frame-to-frame inconsistences that make them unsuitable for precise architectural visualizations. A more accurate, design-focused alternative is the image-to-video workflow: Algorithms like Neural Radiance Fields and the newer 3D Gaussian Splatting can take a handful of building renderings such as orthographic elevations or perspective snapshots and synthesize a smooth orbital walkthrough in seconds, avoiding the lengthy GPU renders of traditional 3D video generation.

7 AI Agents

In June 2023, OpenAI quietly introduced GPT-4 and GPT-3.5 models equipped with function calling capabilities. This event, although not announced on a big stage but merely mentioned in one of OpenAI's changelogs, marked a profound paradigm shift. Previously, LLMs were limited to generating text. At this point, they gained the ability to interact with external systems—thus acting as AI agents—enabling a wide range of real-world applications: from automating daily tasks like managing calendars or sending emails, to querying databases, controlling machines, or integrating with engineering software. Due to the current relevance of these systems, all AI agents mentioned in the book refer to LLM-based agents unless otherwise specified.

AI agents are already among us and can have numerous applications in the field of sustainable buildings. This chapter starts by explaining how AI agents work behind the scenes according to a new, widely accepted standard for their creation. It then explores the agentic "deep research" tool provided by many online LLM services and demonstrates its use in investigative tasks for sustainable projects. Following this, the chapter presents an online platform that offers a no-code method for developing AI agents, illustrating their application to increase productivity in areas like project management of sustainable buildings. The chapter then discusses how to integrate agentic AI into the building planning process as 3D AI assistants, supported by my experience creating an app with ten thousand users that follows this approach, although the app itself is not covered in the book. Finally, the chapter looks ahead to how AI agents are likely to evolve in the building industry.

DOI: 10.1201/9781003506324-7

Introduction to AI Agents and Model Context Protocol

AI agents are autonomous software programs that use tools to accomplish goals. In the context of modern "AI agents", they are built on an LLM that not only generates text but can also autonomously decide to use external tools or software when necessary. The breakthrough is that these external tools are triggered by the LLM itself, which infers from the prompt which tool to call, when to call it, and with what parameters. Therefore, AI agents are more than just software with pre-programmed actions. For example, an app that calculates caloric balance after users input their data and press "submit" then uses an LLM to create a nutrition plan based on that balance, cannot be classified as an AI agent. By contrast, an AI agent could autonomously take the same user's health goals and daily activity updates, decide on its own when to gather new data from a fitness tracker, adjust the nutrition plan accordingly, and send proactive recommendations without being prompted.

The method LLMs use to invoke external tools is not at all mysterious. They receive descriptions and schemas[1] of available tools alongside the user prompt. Using this information, the LLM might generate a response not in natural language, but in a structured format that aligns with a tool's schema. This allows the output to be automatically identified as a "tool call," enabling the execution of the relevant function. Fine-tuning is required for an LLM to create such structured "tool call" responses, which means not all LLMs have the capability to call tools.

The next question to consider is how external tools connect with LLMs, allowing "tool calls" to reach them and their responses to be delivered to the user. While there are various ways to achieve this, the Model Context Protocol (MCP) has been widely recognized as the optimal solution. Introduced by Anthropic, the creators of the Claude LLM, in November 2024, MCP is an open standard designed to link AI agents with external tools and data. It establishes a common schema for passing contextual information to LLMs and exposing tools and capabilities. Furthermore, it clearly separates the client or "host" (the app housing the LLM) from the server (where the actual tools, databases, or files reside). Thanks to this separation and shared schema, third-party service providers can publish standalone MCP servers—even if they do not run AI agents themselves—so developers can "plug in" an MCP server on demand without worrying about the underlying API, database, or filesystem details, much like plugging peripherals into a PC. Figure 1 explains the process flow of an

[1] A schema refers to a structured specification (typically in JSON format). A tool's schema defines the tool's name, description, parameters, and the expected data types and constraints for each parameter.

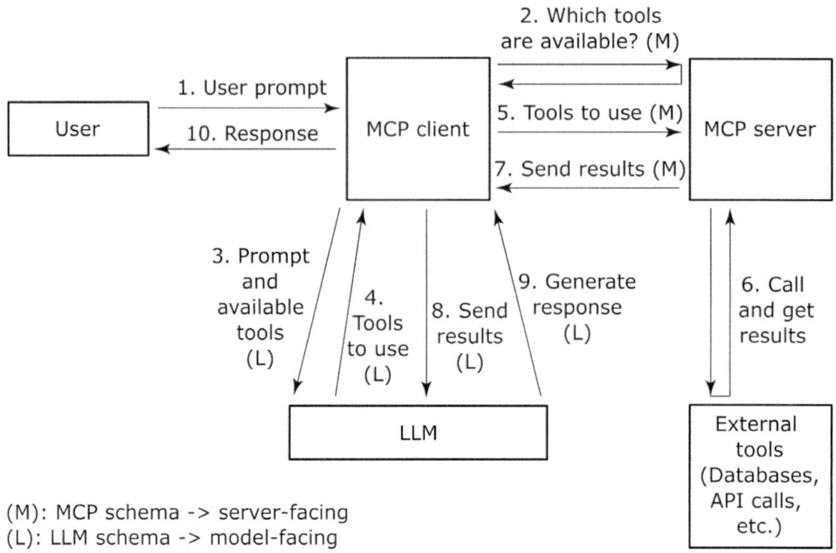

Figure 1. Interaction with an AI agent according to the MCP.

AI agent built on the MCP, detailing how it accesses necessary tools to address the user's request and subsequently generates a response based on that interaction. Additionally, major LLM providers have recently started offering wrappers that translate their own tool schemas to MCP, making it as easy to change the underlying model of the agent as it is to switch external tools and data sources.

Some major cloud LLM providers are already incorporating built-in agents to improve their services. For instance, ChatGPT includes a web-browsing feature the model can invoke by default whenever it needs to search for information released after its training cutoff. Additionally, ChatGPT discreetly employs a calculator tool (or equivalent) to carry out complex mathematical tasks, minimizing the chance of hallucinating incorrect results. Nevertheless, the greatest potential of AI agents typically lies in their use within custom applications.

Following this foundational introduction to AI agents, the subsequent sections will explore different methods for utilizing AI agents during the project phase to enhance sustainability within the building industry. These methods include pre-existing agents that are ready for immediate use and simple, low-code solutions to deploy AI agents. Additionally, the integration of AI agents with robotics holds significant promise for the construction phase, a topic that will be covered in the next chapter.

Deep (Re)search Agents for Innovative Sustainable Planning

The first type of agent discussed in this chapter, due to its accessibility and ease of use, is the deep research agent. These agents are closely connected to the RAG algorithm introduced in Chapter 5, which enables LLMs to generate responses based on custom data sources beyond the model's original training data. RAG has gained significant popularity, and even LLM providers now offer built-in features to search for information in custom documents, perform web searches, and more. However, in certain cases, RAG can be enhanced with an agentic approach, where the LLM actively conducts an iterative search as it gathers information, rather than simply summarizing the top N retrieved data sources to respond to the prompt.

This agentic RAG approach has been offered at several LLM services since the start of 2025 as a so-called "deep research" feature. While the architecture of these tools has been kept secret by the AI corporations, open-source tools like DeepSearcher (ZillizTech 2025) have emerged, replicating the same functionality and outcomes with open code, thus making it possible to understand its agentic architecture. In "deep research," once a user inputs a question, the agent's LLM generates a search query, often rewording the question slightly to enhance the results. It then retrieves relevant documents using embedding search from linked databases or the internet. The LLM assesses these documents and decides whether a subsequent search with a revised query is necessary to gather more valuable information. This iterative process continues until the agent finds the information sufficient, at which point it summarizes it for the user to answer the original question. Nevertheless, standard RAG without agentic functionalities is still highly relevant. While agentic RAG can be very effective for tackling complex queries, it is not always necessary to run such an in-depth, iterative search for every request to a custom database, especially given the extra computation time it requires.

Agentic RAG, or deep research, can become helpful when planning sustainable buildings. With the rising demand for unconventional bio-based materials and reusable structural components, innovation is rapidly advancing in this field. Consequently, it's essential to investigate the availability of new sustainable products for each project. Chapter 5 has already highlighted the potential use of chatbots in this context. However, such investigative tasks often become so intricate that deep research can greatly ease the planning process. For example, timber slabs usually require support from beams or slabs and cannot typically rest directly on columns like flat concrete slabs. Yet, in certain situations, such as misaligned columns complicating beam placement or ventilation duct clashes under the slab, a structural engineer might need to explore recent innovations for a timber point-supported slab. Indeed, some suppliers have recently created such column-slab connections. After examining the product catalog, it is crucial to verify the product certification for compatibility with specific project needs, taking into account geometric constraints, fire resistance, and more. After this, another innovative product would be required to seamlessly connect timber slabs at the edges where timber beams were previously used, which must also be evaluated for project suitability. This is just one of many possible examples that underscore the potential of deep research in sustainable building projects.

Most predefined deep research tools of LLM providers can help in these situations. OpenAI even already provides the option to use their deep research agent with private documents stored in a company's SharePoint cloud, which could be useful, for instance, when references to a specific product used in a privately stored past project are relevant. However, a customized agentic RAG, equipped with a system prompt outlining the office's design approach, a custom document retriever tool managed by the LLM, and even an additional tool to draft and send query emails to product manufacturers, would likely yield even more comprehensive and precise results. There remains considerable room for innovation in agentic RAG.

Automating Project Management with AI Agents and Other Productivity Hacks

Project management is the discipline responsible for planning and coordinating all phases of a project to meet objectives related to time, cost, and quality. In sustainable building projects, this role expands beyond basic tasks like tracking schedules, managing budgets, and communicating with stakeholders. It must also supervise additional processes such as the evaluation of life-cycle impacts, the assessment of energy performance, the procurement of sustainable materials, the application for certifications (such as LEED or BREEAM), and the acquisition of subsidies for eco-friendly buildings. These oversight and coordination duties can be so time-intensive that, for larger projects, building owners often hire external consultants to handle them. By automating certain project management tasks, the overall cost of sustainable projects could be lowered, making them more appealing and accessible to building owners and developers.

There are multiple reasons why an LLM is needed to automate processes related to project management. For example, verifying that all required documentation has been submitted by the planning team by a given deadline requires some level of understanding of the documents themselves; assigning responsibilities during project meetings involves identifying unresolved issues; and creating and maintaining the project timeline depends on a clear understanding of the building owner's goals.

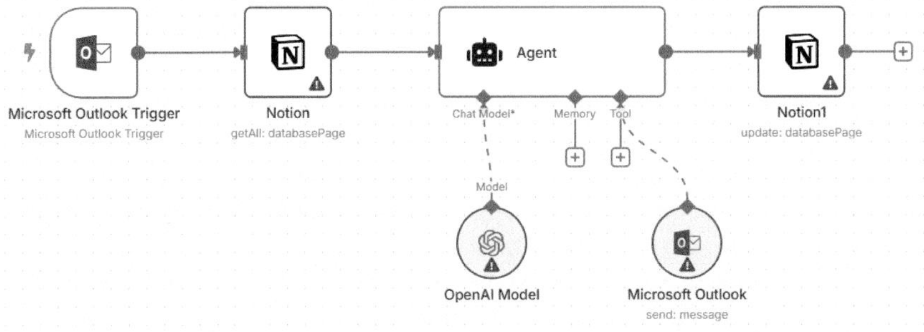

Figure 2. End-to-end n8n workflow for tracking document submissions and notifying stakeholders.

While it is certainly possible to develop AI agents for project management through custom coding, there are also online tools that simplify this process. One such tool is n8n, a workflow automation platform launched in 2019. It features a drag-and-drop, node-based editor—much like visual scripting in Grasshopper3D or ComfyUI—that allows users to connect apps, APIs, and services to automate complex processes. It also includes nodes for configuring AI agents that interact with external tools.

Figure 2 presents a mock-up project management workflow built with n8n, designed to verify that all required documents have been submitted by project stakeholders ahead of a given deadline. The workflow is triggered by an incoming email, after which it queries a Notion database to retrieve the submission status of each planner. An AI agent then reviews these entries, determines whether all documents have been submitted, and drafts either a confirmation email or a polite reminder to those with outstanding submissions. Once the message is prepared, the workflow logs the action back into Notion for documentation and traceability.

Needless to say, AI agents built with n8n have the potential to boost productivity well beyond project management tasks. Using predefined n8n nodes, it is possible to summarize meeting notes from Microsoft Teams, automatically classify incoming emails in Outlook, generate draft reports from Excel data, sync documents between Notion and SharePoint, etc. The possibilities are virtually endless.

Enabling AI Assistants in Building Design

Apple's recent paper (Shojaee et al. 2025), "The Illusion of Thinking," highlights the limited capabilities of LLMs and reasoning LLMs in solving spatial logic puzzles such as the Tower of Hanoi, Checkers Jumping, River Crossing, and Blocks World. These findings are noteworthy because building design can also be seen as a type of spatial logic puzzle: placing columns to support slabs, adjusting their positions to accommodate architectural needs, determining the need for beams, checking for clashes with ventilation or electrical ducts, relocating them if necessary, and so on. Thus, no AI is currently advanced enough to design buildings independently, and human designers' jobs remain secure. However, agentic AI can still play a meaningful role in sustainable building design, primarily as AI assistants.[2] These agents are designed to support and assist humans by performing specific tasks or providing targeted information, rather than operating autonomously. Since design software usually offers APIs, these assistants can effectively interact with the tools professionals already use.

This section introduces an approach that brings AI assistants into Rhino and Grasshopper 3D's computational design ecosystem. As discussed in Chapter 2, computational or algorithmic design has become extended in building practice, allowing teams to explore many alternatives, automate repetitive tasks, model intricate geometries, and minimize human error. However, sharing Grasshopper scripts remains challenging because as a visual program becomes more complex, its intertwined logic can be nearly impossible for others to untangle. The app addresses this issue by letting the creator of a Grasshopper visual script turn it into an AI assistant that others can use. The creator just needs to provide a short explanation of the script and list its input and output parameters. After that, other users can load the AI assistant through a chat interface in Rhino and interact with it directly, without ever needing to deal with the visual programming script itself.

Figure 3 presents a mock-up that illustrates how an AI assistant can help design family houses. To manage the complexity inherent in building design, the assistant is connected to a curated library of floor layout types, so it does not need to generate layouts from scratch. The interaction human-machine becomes therefore a simple process. Initially, the assistant provides design advice if requested. Once the assistant understands the demands of the user regarding the main parameters for the house such as the number of bedrooms, floors, and other requirements, the assistant calls its Grasshopper script to generate a 3D model based on the most suitable layout type. This 3D model can be explored directly in the Rhino viewport. Furthermore, it is also connected to a LCA defined in Grasshopper to calculate and display the building's carbon footprint. Finally, an AI visualization based on the 3D model with the ControlNet approach is provided, giving the user a sense of what the finished house

2 More information on the difference of AI agents and AI assistants on IBM (2024 n.d.).

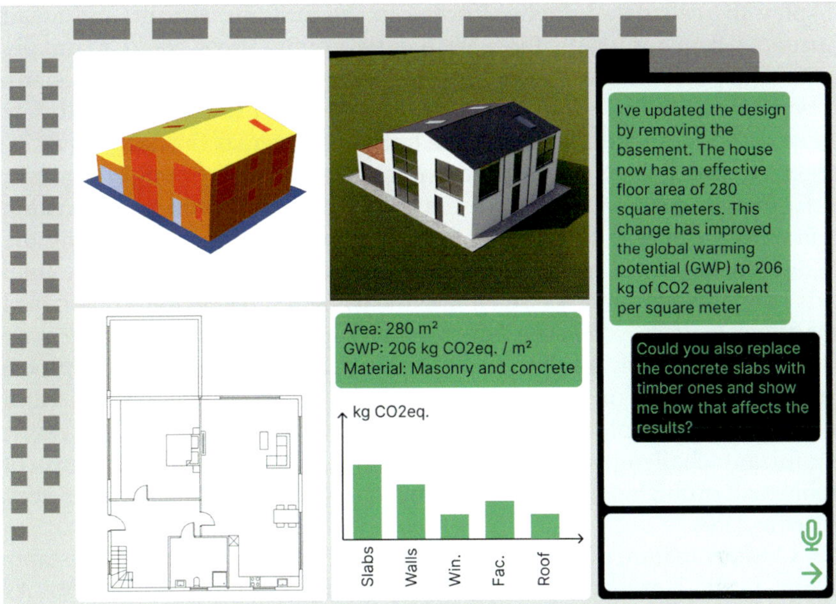

Figure 3. Mock-up of the integration of AI assistants into Rhino.

might look like. After being presented with this design, the user can continue interacting with the assistant to iteratively modify the result, for example, to further reduce the carbon footprint, as illustrated in the mock-up. In addition to the chat interface, users have access to other tabs within the plug-in to select which LLM to interact with, load available assistants, and manage their account details.

The launch of the prototype of such an AI assistants app quickly captured the interest of the computational design community. Over 200 participants attended the initial webinar, and downloads soon reached 1000. However, most users were initially driven by curiosity about AI, with only a handful of early adopters continuing to use the tool throughout the first month. It was from this dedicated group that valuable feedback began to emerge. For example, users expressed a desire for deeper interaction with the geometry produced by the AI assistants, beyond simply zooming, rotating, and examining the models. In response, "handle points" were introduced, enabling users to manipulate the results generated by the assistant more directly. During a second webinar, which drew more than 300 viewers, a case study was presented that combined the app with the environmental analysis tool Ladybug, enabling daylight studies on a building façade. Users could recommend the placement of shading elements themselves or request the assistant's suggestions. In both scenarios, the assistant would model the required shading devices, display the calculated daylight hours in the building atrium, and await further instruction. With the new handle points, users could manually adjust the position of each shading element generated by the AI assistant. It was also extremely interesting to see how the LLM could accurately convert the provided location of the building, given as the name of a city or region, into the actual world coordinates required for sunlight analysis. A further innovation allowed users to query information from existing Rhino 3D models and also from Revit BIM models, by incorporating the "Rhino.Inside.Revit" plug-in into the AI assistant's workflow. The final enhancement explored was, enabling the assistant to generate code-based solutions to user problems beyond the capabilities of the underlying Grasshopper script, similar to how ChatGPT produces code snippets for ad-hoc tasks. For example, the assistant could generate a structural system from a simple sketch and calculate its carbon footprint. However, significant challenges remain in this area, due to the complexity of geometric problems for LLMs and the need for the LLM to master very specific APIs.

One key feature of the app was the ability to interact with AI assistants using voice commands rather than typing prompts manually. As mentioned in Chapter 6, audio-to-text workflows offer clear benefits, since most people speak faster than they type. However, the advantages go even further: users can perform other tasks, such as examining the 3D model, while communicating with the assistants, leading to a significant increase in productivity. An exaggerated example of this multitasking potential is Marvel's superhero Ironman who usually talks to his AI assistant "Jarvis" to activate different widgets of his suit while fighting villains. Back in the real world, the app uses OpenAI's Whisper model, which delivers excellent accuracy in transcribing most voice commands. Still, it is important to note that Whisper is an AI model that transcribes speech in context, processing whole phrases rather than transcribing words one by one. Therefore, it is advisable to speak in complete sentences, rather than using monosyllables, to avoid misunderstandings. Similarly, when referencing for example areas of a building between axes "E" and "G," it is better to use for instance the International Radiotelephony Spelling Alphabet (in this case "Echo" and "Golf") to minimize transcription errors.

A final point to emphasize is that, as previously mentioned, the use of AI assistants or agents in the planning process is largely limited to software environments with extensive APIs, such as Rhino and Grasshopper3D, that enable agents to execute scripts and interact directly with the software. However, many of the programs used in building planning either lack APIs altogether or offer only very limited ones, which significantly restricts the potential for deploying AI agents across the full range of tools used in the industry. An emerging solution to this challenge is illustrated by OpenAI's Operator (OpenAI 2025b) agent, which is based on the so-called computer-using agent (CUA) model. This approach combines GPT-4o's vision capabilities with advanced reasoning, acquired through reinforcement learning, to interact with software in a fundamentally different way: instead of relying on APIs, the agent processes screenshots of websites or applications and interacts with them by controlling the mouse and keyboard, essentially behaving like a human user.

Towards Online AI Agents to Reduce the Carbon Footprint of Basic Buildings

The previous section focused on using AI agents as assistants to planners, since the complexity of holistic building design prevents agents from handling the entire process on their own. However, there is real potential for autonomous AI agents in the design of simple buildings, such as standard industrial halls, storage buildings, or catalogue-based prefabricated homes. These types of buildings depend on a few fundamental geometric parameters, so a single script could be enough to control the whole design, with the AI agent operating it as an external tool. A multi-agent approach is also thinkable, where one agent handles the architecture, another manages the structure, another takes care of MEP, and another runs the LCA and provides feedback, etc. At the end of the process, a human designer would just need to supervise the AI-generated design and approve it for delivery.

Adopting this agentic workflow could significantly enhance the sustainability of the built environment. Currently, modern design firms focused on sustainable planning often avoid taking on basic projects because the lower fees do not justify the effort. However, if an AI agent could manage most of the planning process by directly communicating with clients, requiring human intervention only at the final stage, this scenario would change significantly. Design firms would just need to program an AI agent for a specific type of simple building and launch it online to engage directly with clients. This can be a technically challenging process, but since there are already platforms like the "n8n workflow automation platform" for setting up online AI agents for productivity tools, this approach would simply need to be adapted and introduced to the building industry as well.

Reinforcement Learning

8

Reinforcement learning is a category of ML algorithms where an agent learns by interacting with its environment through trial and error to select a series of actions that achieve a specific objective. Although this area of ML has been somewhat eclipsed by the rise of LLMs, it has nonetheless led to significant scientific advancements. In March 2016, DeepMind's reinforcement learning system, AlphaGo (Silver et al. 2017), defeated world champion Lee Sedol in the game of Go, a feat previously considered impossible for computers. By December 2017, their AlphaZero (Silver et al. 2018) agent taught itself chess (as well as shogi and Go) from the rules alone and outperformed the leading Stockfish engine within 24 hours. In 2019, OpenAI Five triumphed over the world champion team in the intricate multiplayer video game Dota 2 (Berner et al. 2019), showcasing the potential of multi-agent reinforcement learning in a complex, partially observable setting. These accomplishments highlight reinforcement learning's capacity to exceed human expertise in intricate, simulated environments, but its capabilities extend well beyond gaming.

This chapter explores how reinforcement learning can contribute in making our built environment more sustainable in a variety of ways. The theoretical introduction covers various algorithms used to develop self-trained reinforcement learning agents, akin to those developed by DeepMind and OpenAI. The subsequent section, drawn from a chapter of my PhD thesis, examines how these agents can be applied to design building structures utilizing collections of reused elements. Beyond using pure reinforcement learning algorithms for sustainable buildings, merging reinforcement learning with large generative AI models introduces new opportunities. For instance, in robotics, autonomous machines hold substantial promise for assembling sustainable building components, such as timber modules, and even for economically constructing on-site

DOI: 10.1201/9781003506324-8

sustainable systems like masonry vaulted slabs reminiscent of those from a century ago. Hence, the following section investigates how robots can employ reinforcement learning to optimize their movement mechanics for these tasks, while LLMs assist in processing human instructions and VLMs enhance their understanding of the environment. The concluding section discusses the fine-tuning of large models using reinforcement learning, which provides a significant advantage in applications related to sustainable building design. This fine-tuning approach is essential for developing efficient LLMs such as the recent and highly regarded DeepSeek model and is expected to become increasingly relevant in the years to come.

Reinforcement Learning Agents

Reinforcement learning agents learn to make decisions in step-by-step environments, aiming for the best possible results over the long run. The kinds of problems they solve are known as Markov decision processes (MDPs). In these situations, the agent only needs to consider the current state to decide what to do next because all the important information is already present in it.

MDPs play a key role in video games and board games, since each new state depends on the previous one and the latest action taken. However, MDPs are not limited to games. Plenty of real-world situations can also be modeled this way. For instance, the MuZero agent[1]—an evolution of the chess-playing AlphaZero—has been successfully applied to challenges like video compression. This is possible because a video can be represented as a sequence of frames, forming an MDP. The success of MuZero demonstrates that reinforcement learning agents developed for games have practical value beyond entertainment, with real-world applications emerging more quickly than many might expect. OpenAI is another good example: before creating ChatGPT, they released "OpenAI Gym," a collection of simple, arcade-style environments that let developers and researchers from all over the world experiment with and improve their agents. Some of OpenAI's algorithms developed in these environments, like Proximal Policy Optimization (PPO), have become especially influential. PPO is for instance very popular in the field of robotics and it also forms the backbone of the reinforcement learning from human feedback approach used to fine-tune LLMs such as ChatGPT.

[1] MuZero is an algorithm that learns a model of the game's dynamics (including rules and rewards) and combines planning with learned representations, enabling it to handle environments without known rules.

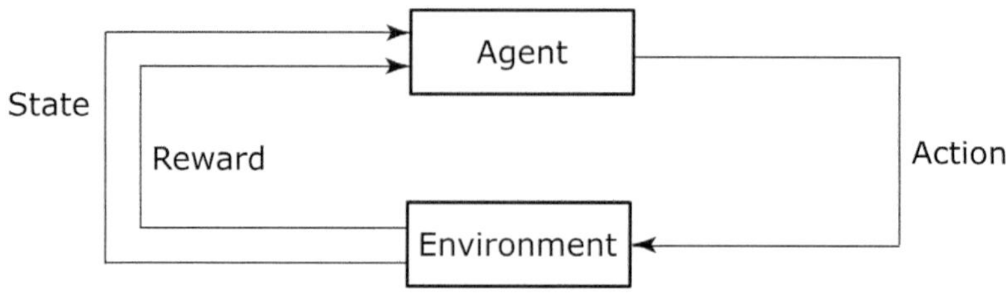

Figure 1. The reinforcement learning agent-environment interaction loop.

Figure 1 shows how the training loop works in reinforcement learning. At each step, the agent looks at the current situation or state and chooses an action. The environment reacts to this action by updating the state and providing feedback in the form of a "reward" that tells the agent how good or bad its choice was. When an episode[2] or game ends, the agent might get an extra reward if it reached its goal. Then, a new episode begins and the process repeats. The agent uses this feedback accumulated over the episodes to train its model—often a neural network—so it can make better decisions over time and maximize the total reward it earns in each episode. In the early stages of training, however, the agent often makes random moves, disregarding the model, to explore the environment and find out which strategies lead to higher rewards. Once training is done, the agent can be put back into the environment, but now it focuses on performing as well as possible instead of learning. In this way, the agent exhibits a unique form of autonomy: it not only makes its own decisions within the environment but also generates its own training data through these interactions, continuously improving its model based on the feedback it receives.

Reinforcement learning algorithms are broadly classified by whether they operate with discrete or continuous states, and whether their actions are discrete or continuous. For instance, the Monte Carlo Tree Search (MCTS) algorithm is specifically designed for discrete states and actions, which is why it is commonly used in agents for board games and similar tasks. The training process for MCTS follows the steps illustrated in Figure 2. It begins with navigating the tree from the root node until a node that is not fully expanded, meaning an environment state with unexplored actions, is reached. From there, a new child node is created by taking an action that has not yet been tried in that state. Next, the algorithm simulates random actions from this point, discovering successive nodes until it reaches a terminal state.[3] Finally, the reward associated with this

[2] An episode is a complete sequence of states, actions, and rewards, typically ending in a terminal state or goal.

[3] A terminal state is a state in which no further actions are possible, marking the end of an episode.

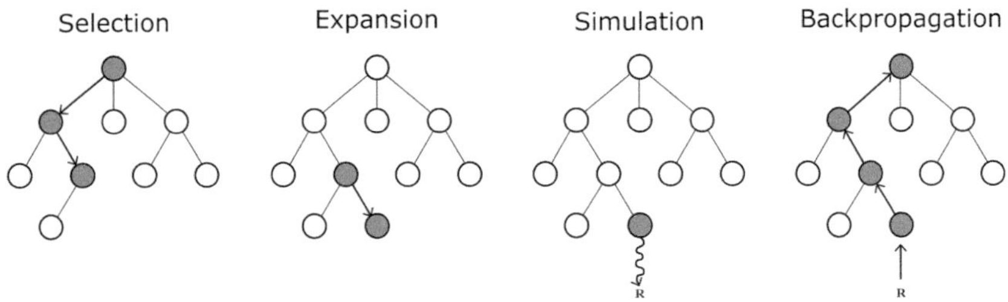

Figure 2. Monte Carlo tree search algorithm.

terminal state is calculated and passed back up through the tree, helping to determine which nodes should be prioritized in future selections.

The standard MCTS algorithm operates without the need for neural networks. It works by simulating numerous possible sequences of actions, gradually expanding the search tree until it is well-developed. Once the tree has grown, the MCTS agent chooses its actions based on the nodes with the highest scores. In contrast, the AlphaZero algorithm, which is based on MCTS, introduces a powerful twist: it implements a CNN that estimates the value of a given state. This integration allows the search process to be stopped early, relying on the neural network's evaluation rather than playing out each simulation to the end. By blending these approaches, AlphaZero can tackle games with a vast number of possible outcomes, such as chess or Go, which would be far too complex to solve with MCTS alone.

Another widely used reinforcement learning algorithm is the Deep Q-Network (DQN) (Mnih et al. 2015). DQN employs a neural network to estimate what is known as the "Q-value": an expected measure of long-term reward after each possible action in a given state. Each interaction produces an experience that consists of a state, an action, the reward received, and the resulting next state. Experiences are stored in what is called a replay buffer. During training, the algorithm randomly samples from this buffer. Then, it adjusts its predictions of the Q-values to be closer to the sum of the reward it just got and its own estimate of the Q-value from the next state. By repeating this process, the neural network's predictions become increasingly accurate. Over time, the agent learns to select actions in each state that maximize its expected long-term reward.

Figure 3 (left) shows a DQN agent playing the classic Snake game. In this example, the agent's CNN processes the game board, represented as a 5 × 5 pixel image where empty cells, food, the snake's head, and its body are each encoded with different colors. Based on this input, the trained agent decides at every step whether to move straight, turn left, or turn right, always aiming to maximize its total score. Interestingly, the best move is not always the most direct path to the food because sometimes avoiding

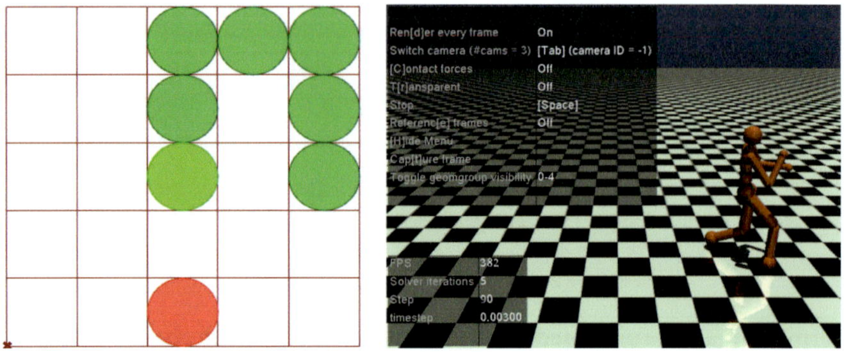

Figure 3. Reinforcement learning agents applied to two different environments: the snake game (left) and the MuJoCo Humanoid environment (right).

an immediate reward is necessary to prevent the snake from trapping itself. Since the number of possible situations is too large for hand-coded strategies, the agent's ability to generalize from past experience becomes crucial. After many games, the agent learns to reliably predict which actions will yield the highest final score, even when faced with situations it has never exactly seen before.

Unlike DQN, which is limited to discrete actions, some reinforcement learning algorithms are designed to work with continuous action spaces. This capability is essential in fields like robotics, where a robot may need to adjust its movements with precision, rather than simply choosing between "left" or "right." One of the most widely used algorithms for such cases is the previously introduced PPO. It employs a single neural network with two outputs, often called "heads." The "actor" head generates the parameters (such as mean and standard deviation) for a probability distribution over possible actions, allowing the agent to select actions from a continuous range. The "critic" head evaluates how favorable a given situation is by predicting the total future reward from that state, thus guiding learning by providing feedback on the quality of those actions. During training, PPO constrains how much the agent's strategy, or policy, can change with each update. This incremental approach helps maintain stable learning by avoiding drastic shifts in the agent's behavior.

Figure 3 (right) depicts a PPO agent operating in the Multi-Joint dynamics with Contact (MuJoCo) Humanoid environment: a physics-based simulation in which a robot must control the torque applied to each joint at every time step to walk and maintain its balance. In this setting, the action space is both high-dimensional and continuous, making it infeasible to manually design rules for every possible scenario. Instead, the PPO agent learns an effective movement policy from scratch by experimenting with different actions and seeking to maximize its expected cumulative reward. Here, the actor-critic framework is particularly valuable: the actor explores possible movements, while the critic evaluates them, enabling the agent to gradually improve its performance even in such a complex, dynamic environment.

Enabling Circular Economy in Building Design

Chapter 2 already discussed the environmental benefits of introducing reuse in construction. More and more modern buildings are adopting a design-for-disassembly strategy, enabling them to be easily taken apart at the end of their lifecycle, which could promote the reuse of their components. Nonetheless, the actual process of reusing these parts once a building has been dismantled remains a complicated issue that requires greater focus. Additionally, although this topic is beyond the scope of this book, there is a need for stronger political support to promote this progressive design approach, to encourage standardized formats for building components, and to support public databases where reclaimed components can be sourced, thus making widespread reuse more feasible. Despite these challenges, some pioneering projects are starting to show what is possible. For example, the cultural building "Buitenplaats Brienenoord" in Rotterdam reused steel columns and trusses from a former hall as roof girders and bracing elements, although for different structural purposes (Schiferli and Winkler 2022). Similarly, the Mehr.WERT.Pavillon in Heilbronn is supported by a steel tree structure made from reused steel pipes (Heisel et al. 2019). While these projects were probably intended to showcase the potential of reusing building components, they also highlight just how complex this process can be, and how much planning it requires. If architecture based on reused components is to become the rule rather than the exception, AI will be essential in supporting designers as they plan how to best incorporate these elements into new buildings. The examples presented in this section illustrate how reinforcement learning agents can be applied to this challenge. The first example is based on a real educational project that used a stock of bamboo poles. The second focuses on using reclaimed steel and timber beams to create new truss structures, and includes a comparison demonstrating how reinforcement learning can outperform other algorithms and even manual design for these tasks. As discussed in Chapter 7, LLMs are not yet suitable for complex spatial planning and are therefore not included in this discussion, but it will be interesting to see how this technology develops in the coming years.

The bamboo hall at the Maya Kaqchikel school in Guatemala, shown in Figure 4, was designed and built in 2019 by a team from the Berlin University of the Arts and the Technical University of Berlin, working together with local architects and construction workers. One of the main challenges in this project was figuring out which bamboo poles to use

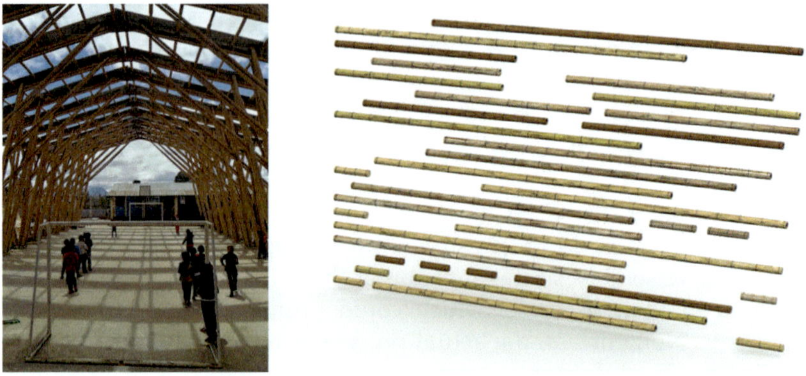

Figure 4. Photo of the bamboo hall of Maya Kaqchikel, Guatemala (© Michel Schmeck, left), and a visualization of bamboo selection by the DQN agent for one of the structural frames (right).

for each part of the structure. When building with bamboo, it is crucial to have a detailed catalogue of all available poles. This means not just listing their length and diameter, but also the exact position of every node along each culm. This is especially important because, according to the applicable structural code, only two bolted connections are allowed within a single internode. With the particular bamboo species used in this project (Asper), the close and irregular spacing of nodes added yet another layer of complexity to selecting the right segment for each structural member. To tackle this challenge in future buildings, the project was used as a case study to train a reinforcement agent. The goal was to optimize the selection of bamboo poles from the existing stock because automating this process offers several advantages. First, it can speed up construction by quickly identifying the best poles for each structural member. Second, by efficiently using the available poles across different parts of the structure, the system can help reduce the total number of bamboo poles required. This means fewer poles have to be transported from distant plantations, which in turn significantly lowers the project's carbon footprint.

The problem of the bamboo allocation is formulated as a MDP, where at each step, a bamboo pole is selected for a structural member. The main reason for using a MCTS agent instead of a DQN agent is that the number of compatible poles for each structural member, and therefore the number of possible actions, varies at every step. The stock of bamboo poles in the reinforcement learning environment was defined using actual data from the bamboo available on the construction site in Guatemala. The reward system is designed to encourage efficiency. The shorter the leftover segment of a pole after it has been used, the higher the intermediate reward at each step. If the agent successfully assigns compatible bamboo segments to all structural members, taking into account both the required structural capacity and the location of nodes for bolted connections, it receives an additional reward at the end of the episode. Figure 4 shows the remaining bamboo stock after the trained MCTS agent selects suitable segments for one of the hall's frames. It can be noticed how the agent has learned to extract the smaller members of the frame from a single bamboo pole, leaving the longer poles untouched so they can potentially be used for larger columns or rafters later. This approach not only reduces waste but also minimizes the total number of poles that need to be transported to the construction site.

The next case study focuses on the configuration of trusses made from reclaimed steel and timber elements. Since trusses are common in architecture and can be assembled from individual reused beams of various origins, this problem has already been addressed in several scientific publications proposing different algorithms and approaches to tackle the challenge. Figure 5 presents the results from vii, comparing three different approaches to generating truss structures on a rectangular grid using reused elements. Among these, the DQN agent achieved the lowest embodied carbon. In this approach, the MDP consists of a DQN agent "traveling" through the grid and deciding at each point whether to place a truss node. After training for 1,000 episodes, the agent learns the optimal policy for generating truss structures based on the given stock of elements. If a different stock of elements or a different rectangular grid is to be studied, the DQN agent must be retrained. Nevertheless, the definition of the MDP would require only minimal adaptation for these new scenarios.

The optimal structural configuration, in terms of carbon footprint or global warming potential, strongly depends on the available stock of materials. In this case, the environmentally friendly timber beams, which are slightly less strong than the steel ones, could be used by the agent at their full length for members with lower forces, such as the end beam of the bottom chord and the truss diagonals. The lower chord of the truss, which is under tension, could be completed by the agent using the available slender steel profile. The agent had to use the more massive steel profile for the upper chord to avoid buckling failure due to the high compressive forces. Regarding the relatively low height of the truss, it leads to high tension forces in the chords. However, since these forces in the lower chord can be carried by the available slender steel profile, the reduced length of the lower chord leads to a reduction in material use.

In contrast, the genetic algorithm does not truly converge to an optimal solution even after ten thousand iterations, as it fails to align all the nodes of the top chord along the same horizontal axis. The original outcome of the genetic algorithm is shown with a low level of transparency, overlaid on a manually adapted version of the same result. Among the three approaches, this solution still results in the highest embodied carbon. The

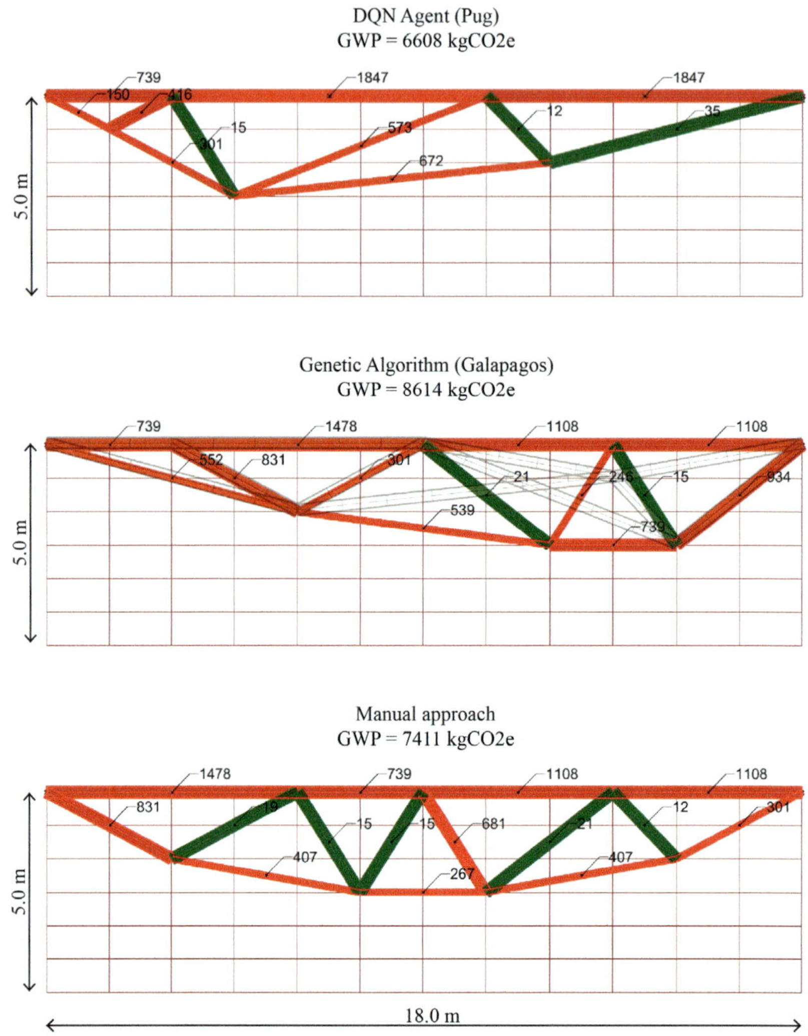

Figure 5. Comparison of truss structures generated from re-used steel and timber elements. The DQN agent achieves better results than both the genetic algorithm Galapagos and manual design methods.

inefficiency of the genetic algorithm is most likely due to discontinuous problem landscape and the relatively high number of genes involved.

Finally, the manual approach involved selecting a regular truss configuration with the same height as the optimal result produced by the DQN agent. However, the embodied carbon remained significantly higher than that of the structure generated by the DQN agent. This highlights the importance of using AI agents in designing building structures, especially within the context of a circular economy.

AI-Powered Robotics in Construction

Robots are playing an increasingly important role in construction, especially in prefabrication. By moving much of the work into controlled factory environments, companies can boost efficiency, quality, and safety (Long 2020). This shift is closely tied to sustainability: factories use materials more efficiently, run on electricity rather than diesel, and reduce waste thanks to optimized processes. Furthermore, robotics are essential for working with some low-carbon building materials. For example, timber-frame components are usually factory-made for faster assembly. Modular construction with timber, such as pre-assembled bathroom or kitchen pods, is also becoming more common because it allows for quick installation on site. Even load-bearing walls made of rammed earth are now often produced in factories, where robots mechanically press them to achieve the required strength and consistency. Therefore, incorporating robotics into these workflows can significantly improve the sustainability of the building industry. Beyond traditional prefabrication, robotic 3D printing makes it possible to create low-carbon structures, particularly when using earth-based materials instead of concrete (Curth et al. 2024). Finally, robots deployed on construction sites can also drive sustainable innovation. For example, there are prototypes of robotic arms that can economically build material-efficient vaulted masonry slabs by automating traditional building techniques (Gengnagel and Brechenmacher 2023).

Consequently, the more autonomous robots become, the more sustainable the building industry can become. AI can play a central role in this process. Construction environments are highly dynamic and uncertain, so it is unrealistic to expect robots to be fully functional with only hand-coded algorithms. AI, on the other hand, can be used to train robots in simulated construction environments, enabling them to use inference and adapt to real-world situations when deployed in assembly factories and on construction sites. Additionally, recent advancements in multimodal sensors can further enhance robot autonomy in construction (Duan et al. 2022). Multimodal sensors are devices or systems that combine data from multiple types of sensors, such as cameras, microphones, LiDAR or radar, tactile sensors, GPS, temperature sensors, and more. By providing a richer understanding of their environment and enabling more effective AI training through a broader range of data points, these sensors lead to more autonomous AI-powered robots

Reinforcement learning has become increasingly popular in robotics research over the past decade. Simulators like the MuJoCo physics engine, as previously mentioned, are widely used to model robots with joints and moving parts, allowing researchers to safely and efficiently test ideas in virtual environments. These tools are suitable for a wide range of robot types, from humanoids to robotic arms, and help teams develop new agents before deploying them on real machines. A valid question is whether these "simulated" agents can perform just as well in the real world. Encouragingly, there has been strong progress: for example, Nvidia recently introduced its AI-powered humanoid robot (Project GR00T), which learned complex movements such as walking and picking up objects entirely through training in simulation. In construction specifically, Zhu et al. developed agents using reinforcement learning algorithms like DQN and PPO to control cranes for assembly tasks and verified their methods in large-scale experiments that closely resembled real construction sites (Zhu et al. 2023). Beyond research, some contractors are already using semi-autonomous cranes. Kangkan et al. have further explored the application of multi-agent robots based on the PPO algorithm, enabling several agents to coordinate and take on different human-related tasks on construction sites (Duan et al. 2023). The Neom mega-project is also employing multiple autonomous robots for various tasks, such as the placement of steel rebars or the transport of materials (India Sea Trade News 2024). Waymo, famous for developing robotaxis, is now automating construction sites with self-driving dirt diggers (Ohnsman 2025). Ultimately, reinforcement learning still holds enormous potential for automating construction sites and assembly factories: from robots that cut timber with less waste, to smarter 3D printers, to fleets of machines that assemble modular buildings with high precision, and much more.

Beyond reinforcement learning, VLMs and LLMs are increasingly present in robotics, providing high-level reasoning and planning capabilities. For example, DeepMind has demonstrated an approach where a VLM is presented with frames from videos of construction tasks (Ma et al. 2024), enabling it to estimate task progress in real scenarios simply by analyzing visual data and textual prompts. This ability allows robots to assess and adapt to unfamiliar situations while performing various tasks. Additionally, DeepMind developed a fine-tuned VLM called "Robotics Transformer 2," which maps visual observations and abstract commands directly to precise robot actions (Zitkovich et al. 2023). Meanwhile, LLMs assist robotic systems by parsing high-level language instructions into concrete task representations, using internal reasoning to break down complex objectives into sequential subtasks, and generating executable code or control policies for specific operations.

Last but not least, combining different AI models creates powerful synergies in robotics. For instance, LLMs can help decompose complex tasks into subtasks and even serve as reward models to evaluate task completion in reinforcement learning scenarios. At the same time, supervised learning remains essential for specific perception tasks such as object detection and classification. By integrating these AI approaches, robotic systems can achieve greater autonomy and flexibility.

Enhancing Large Language Models with Reinforcement Learning

Reinforcement learning has become a key technique for fine-tuning LLMs. This method gained traction with models like WebGPT in 2021 and InstructGPT in early 2022 (Ouyang et al. 2022), and was later used to train GPT-3.5, the series behind the first ChatGPT release in late 2022. These models used RLHF, as described in Chapter 5. RLHF comes after an initial supervised fine-tuning phase and generally involves two main steps. First, human annotators rank several candidate responses from the model, focusing on qualities like helpfulness, truthfulness, and politeness. This ranking data is then used to train a separate reward model through supervised learning. Second, the LLM itself is further fine-tuned using a reinforcement learning loop that is usually PPO. In this setup, the environment is the conversation, the state is the sequence of tokens generated so far, and each action is the model choosing the next token from its vocabulary. PPO aims to maximize the score given by the reward model for the whole generated response, while keeping changes to the model's parameters small for stability. After enough training episodes, the LLM's responses are significantly more aligned with human behavior. For example, the model is more likely to provide accurate instructions for a life-cycle assessment, instead of inventing steps; it can better spot greenwashing when searching for information on sustainable products; and much more.

Although RLHF substantially aligns LLMs with human preferences, models still struggle with multi-step reasoning tasks such as coding, math, and logic problems that require thought processes. To address this, OpenAI introduced a new approach called "large-scale reinforcement learning" that they applied it to the first public reasoning model, OpenAI o1, released in September 2024 (OpenAI 2024). This method builds on the chain-of-thought paradigm that's been popular since 2022, where LLMs perform better on reasoning questions if they're prompted to explain their reasoning step by step before giving a final answer. With large-scale reinforcement learning, the model is fine-tuned not just to produce final answers, but also to generate and evaluate its own intermediate reasoning steps.

Unlike RLHF, large-scale reinforcement learning drops human-generated rewards in favor of automatic rewards, such as checking whether answers to programming, math or logic problems are correct. Fully automating rewards, rather than relying on a reward model constrained by the limited quantity of human-annotated data, allows far more compute to be devoted to the reinforcement-learning phase, enabling deeper optimization. Inspired by these advances, DeepSeek open-sourced R1-Zero in January 2025. This model was trained exclusively via reinforcement learning, with no supervised fine-tuning - much like AlphaZero, which learned entirely through self-play reinforcement learning, unlike its predecessor AlphaGo, which first used supervised learning to bootstrap its Go knowledge. Its follow-up, DeepSeek-R1 (Guo et al. 2025), adds a short supervised phase, but otherwise relies on a new Grouped Relative Policy Optimization (GRPO) algorithm, which eliminates the need for the separate value network used in PPO. DeepSeek-R1 also uses a "mixture-of-experts" model architecture, where each token is routed only through the relevant expert sub-networks. These technical breakthroughs, combined with DeepSeek-R1's open-source release, quickly made DeepSeek's mobile app the top download on Apple's US App Store and since then, most LLM providers have made reasoning models their flagship services. For the building industry, reasoning models present wide-ranging opportunities: they can help designers interpret complex regulations and standards, propose more sustainable design options, and evaluate different alternatives to highlight the most sustainable and suitable choice for a project.

In summary, building on these breakthroughs, reinforcement learning has become a go-to method for fine-tuning generative AI beyond plain LLMs. For example, the browser-navigating agents in Chapter 7 use reinforcement learning to acquire the step-by-step reasoning needed to handle complex web interfaces. Multimodal AI benefits too: VLMs aligned with RLHF generate more accurate captions in text-to-image tasks. While it remains nontrivial for individual users to set up reinforcement learning-based fine-tuning, these advances have already driven noticeable improvements in consumer-facing generative AI products.

Conclusions

9

Insights Gained Along the Journey

The journey of this book now comes to an end. It began in Chapter 1 with the realization that the building industry's impact on climate change is significantly greater than that of artificial intelligence. Public perception may suggest otherwise, as it is shaped by frequent media discussions about AI's energy consumption, while the overall impact of the building industry receives far less attention. Chapter 2 continued by examining how to measure the sustainability and carbon footprint of buildings, emphasizing the crucial role of the building structure in this assessment. Since the purpose of the book is to explore pathways toward sustainable building structures, it was first necessary to define what sustainable structures are and to present common strategies for achieving sustainability in building projects—strategies that may be further enhanced through the application of AI. Chapter 3 provided the final step before exploring the diverse and rapidly evolving field of AI. It was important to review the current state of digitalization within the building industry to later consider how AI technologies could be integrated, much like fitting a plug into a compatible socket. After these foundational stages, the book went on to explore the different AI methodologies, with the goal of understanding them and identifying opportunities that could help move the industry toward more sustainable building projects.

The first stage of AI, presented in Chapter 4, introduced traditional ML approaches that involve training models for specific purposes using large datasets from the relevant field. Despite the challenges inherent in this process, numerous ML-based technologies were identified that can enhance the sustainability of buildings, ranging from the earliest conceptual design stages to extending the structural life cycle and even supporting

DOI: 10.1201/9781003506324-9

renovations. The central theme of this chapter was validation. While these tools offer significant potential, it is essential to thoroughly assess whether an AI model has been properly trained to make meaningful predictions based on its training data, and whether the data itself is reliable.

After establishing a foundation in traditional ML, the journey moved into the realm of generative AI, beginning with LLMs in Chapter 5. Since 2022, chatbots powered by GPT technology have become omnipresent in daily life, though their impact on the building industry has, so far, been limited. The chapter emphasized that the positive effects of this methodology extend well beyond simple task automation, such as generating sustainability reports, to include substantial productivity gains. For example, sourcing information about sustainable materials, researching previous office projects, or writing scripts for design software are just some of the many possibilities that help ensure sustainable building projects can meet the standard deadlines and budgets typically expected by clients. The chapter also explored various methods for identifying hallucinations produced by LLMs, arising from gaps in their training data or the need for more advanced reasoning, and provided strategies for avoiding these issues.

The next stage of the journey explored multimodal generative AI. Chapter 6 examined generative AI models that enable a wide range of workflows beyond standard text-to-text LLMs. For instance, text-to-image generation using diffusion models was shown to support architects in presenting and discussing sustainable projects with clients. However, to ensure that the generated visualizations accurately reflect the anticipated final building, a greater degree of control is often required, something that image-to-image workflows are better suited to provide. The reverse workflow, image-to-text, has also become possible thanks to VLMs. The chapter assessed how proficient AI is at analyzing the built environment by examining existing buildings, automating the evaluation of retrofitting opportunities to help extend building life cycles, and interpreting documents and drawings to streamline renovation planning. Other workflows discussed, such as text-to-3D, audio-to-text, and text-to-video, are based on rapidly evolving technologies and are likely to play an increasingly prominent role in the building industry in the near future.

At this stage, the book's journey had made it clear that the potential of generative AI in the building industry is like an iceberg: what is visible are the straightforward interactions with ChatGPT and other LLM-based chat services that most people are familiar with, while much more lies beneath the surface. Chapter 7 continued the exploration by focusing on AI agents that integrate LLMs with external tools they can access as needed. AI agents are already spreading rapidly across IT-related fields and office software, especially with the advent of MCP. The chapter discussed how these agents can also be applied to improve productivity in various project management areas related to sustainable building projects, as well as to automate specific tasks within building design software. Although there is not yet any AI approach that enables agents to fully understand and design complex buildings independently, the chapter demonstrated how online AI agents connected to parametric design scripts can already interact directly with potential clients for smaller, standardized building types. These agents can advise clients and create preliminary sustainable designs based on client requirements, making sustainable buildings possible for simple projects with limited budgets that otherwise might not have had access to such expertise.

Chapter 8 concluded the review of AI by discussing reinforcement learning. The chapter demonstrated how self-trained reinforcement learning agents, which are well known for their roles in several groundbreaking scientific achievements, can be applied to sustainable buildings. This is especially relevant in the challenging task of designing structures using reclaimed components. The chapter also explained how the powerful and continually evolving synergy between generative AI and reinforcement learning is appearing in areas such as construction robotics. This synergy is also driving significant progress in the fine-tuning of state-of-the-art generative AI models, which further supports their integration into the building industry.

Ensuring Safe Implementation of AI in the Building Industry

The key takeaway from this journey is that implementing AI in the building industry holds significant potential for positive environmental impact and can contribute to reversing climate change. However, it is essential to recognize and address the inherent risks associated with AI to ensure its safe and responsible integration into the industry.

First, this book has outlined many opportunities for AI to enhance the sustainability of our built environment. If these opportunities are ignored and AI is simply used to reinforce conventional design and construction practices, it becomes difficult to justify its adoption beyond a narrow, capitalist perspective focused on maximizing profits and minimizing losses. There is therefore a risk that AI could become merely a tool for increasing productivity in projects that may ultimately exacerbate climate change. Instead, AI should serve as a catalyst for the sustainable transformation of the building industry.

Regarding the impact of AI in workplaces, this book has shown that AI is currently best suited to serve as an assistant to professionals in the building industry. Due to the geometrical, spatial, and physical nature of many tasks in this field, there are relatively few jobs that AI could fully automate. As a result, mass layoffs like those seen in the IT sector are not expected. However, AI will still disrupt the building industry. Individuals, companies, and even countries that embrace AI will gain a greater share of the market compared to those who do not. For this reason, the implementation of AI is no longer optional.

Another risk associated with AI, common across many industries, is ensuring data privacy. This concern is especially relevant in the building industry, where projects are inherently confidential and offices routinely handle sensitive information such as budgets, prices, and processes. If such information were to become publicly accessible, it could give competitors an unfair advantage. Nevertheless, this book has highlighted various techniques that allow AI models to work effectively with private data without compromising its privacy. These include RAG methods that keep private data separate from public training data, deploying private fine-tuned versions of public models, using enterprise subscriptions to public LLM services that do not use customer data for training, and locally deploying AI models that do not have access beyond the office's local network.

Last but not least, the ethics of AI algorithms must be taken into account. In the months following the emergence of generative AI in 2022 and 2023, there was intense debate over whether training these algorithms on copyrighted material, such as textbooks and artworks, was ethical or even legal. On the legal front, a key US court ruling has clarified that using lawfully obtained copyrighted works to train AI models can qualify as fair use under US copyright law.[1] However, the court emphasized that training on pirated content is not protected by fair use. Another concern involves deepfake videos and fabricated news articles, which can now be easily produced using state of the art AI tools. As discussed in Chapter 6, AI generated images and videos can be used to greenwash designs that are far from being sustainable. For this reason, clients and all project stakeholders must develop an awareness of these risks. This begins with understanding what constitutes a sustainable building, how AI algorithms operate, and how to identify fraudulent documentation.

All these approaches demonstrate that the potential risks of AI for the building industry can be mitigated within the sector itself, without requiring policymakers to regulate AI development at the foundational model level. Overregulation at this level would greatly restrict progress, especially compared to countries that take a more liberal approach to AI, and could prevent society from benefiting from the scientific advances AI enables, including the sustainable transformation of the building industry

[1] In a landmark decision in Bartz v. Anthropic, Judge William Alsup of the US District Court for the Northern District of California ruled in June 2025 that such use was "quintessentially transformative" and thus legally permissible.

discussed in this book. Instead, regulating AI at the application level is a more pragmatic way to prevent the risks posed by unethical algorithms, while still allowing for scientific and competitive progress.

A Positive Outlook

The steady progress occurring across multiple areas of AI has been repeatedly emphasized throughout this book. Major corporations are investing significant resources in generative AI to ensure this: from the infrastructure needed to train large AI models (Mann 2024) to acquiring the talent required to develop these complex techniques (Langley 2025). While AI is advancing steadily towards Artificial General Intelligence (AGI), the timeline for achieving it and the full scope of its societal and economic implications remains highly uncertain,[2] making any predictions about this breakthrough necessarily speculative. While the horizon for AGI remains uncertain, we can draw far more concrete conclusions from current trends. One of the most rapidly evolving fields is the fine-tuning of LLMs through large-scale reinforcement learning, as described in Chapter 8, which is producing highly advanced reasoning models. These are now commonly used in applications such as coding AI agents, where many professional software developers pay as much as 200$ per month for these tools because they see clear and direct returns from their use (Brown 2025). The impact AI has already had on the IT sector is undeniable.

Nevertheless, the growing use of AI models will not necessarily result in a proportional increase in energy consumption. There are several indications that AI is becoming more scalable, both in terms of energy use and, consequently, in terms of cost. For example, Andrej Karpathy observed while training the open source nanoGPT model that the process had become orders of magnitude cheaper and faster compared to when GPT-2 was originally trained (Karpathy 2025b). Sam Altman has also noted that, as datacenter production becomes automated, the cost of AI should eventually converge to the cost of electricity (Altman 2017).

Regarding the adoption of AI in the building industry, progress is certainly occurring at a slower pace than in the IT sector. On the one hand,

[2] AGI refers to AI systems capable of human-level performance across any cognitive task far beyond the domain-specific powers of today's models. While major strides, like the release of GPT5, mark "significant steps" toward AGI, they still fall short of autonomous, continuous learning and broad adaptability. AGI promises transformative potential: from breakthroughs in healthcare, the environment, and scientific discovery to radical shifts in the nature of work and economic systems. Yet it also raises profound ethical, societal, and safety concerns including job displacement, concentration of power, and even existential risk.

buildings present a greater challenge for AI due to the physical and spatial complexities of construction projects and the more limited availability of relevant training data. On the other hand, the building industry requires a comparable ecosystem to that of the IT sector: one composed of numerous AI startups willing to challenge established workflows, venture capitalists prepared to invest in these innovations, and, just as importantly, a sufficient number of early adopters who can help implement AI in a sustainable way.

References

Altman, S. 2017. The gentle singularity. Sam Altman's Blog, 13 May. Available at: https://blog.samaltman.com/the-gentle-singularity (accessed 9 August 2025).

Apellániz, D. 2024. A parametric and machine learning-based approach to implement life-cycle assessment in the early design stages. Doctoral thesis, Universität der Künste Berlin, 183 pp. doi:10.25624/kuenste-2346.

Apellániz, D. and R. Vierlinger. 2022. Enhancing structural design with a parametric FEM toolbox. Steel Construction 15(3): 188–195. doi:10.1002/stco.202200004.

Apellániz, D., B. Pettersson and C. Gengnagel. 2023. A flexible reinforcement learning framework to implement cradle-to-cradle in early design stages. *In*: Gengnagel, C., O. Baverel, G. Betti, M. Popescu, M.R. Thomsen and J. Wurm (eds.). Towards radical regeneration. DMS 2022. Springer, Cham. doi:10.1007/978-3-031-13249-0_1.

Apellániz, D., C. Groth, H. Alpermann and C. Gengnagel. 2019. Strengthening of the Transfer Floor of Berlin's Steglitzer Kreisel. Proceedings of IASS Annual Symposia 2019(1): 1–8.

Apellániz, D., P. Pasanen and C. Gengnagel. 2021. A holistic and parametric approach for life cycle assessment in the early design stages. In: Proceedings of the Symposium on Simulation for Architecture and Urban Design (SimAUD), April 2021.

Bennett, P. 2023. Climate change is costing the world $16 million per hour: study. World Economic Forum, 12 October. Available at: https://www.weforum.org/stories/2023/10/climate-loss-and-damage-cost-16-million-per-hour (accessed 9 August 2025).

Berner, C., G. Brockman, B. Chan, V. Cheung, P. Dębiak, C. Dennison et al. 2019. Dota 2 with large scale deep reinforcement learning. arXiv preprint arXiv:1912.06680.

Billington, D.P. 1983. The tower and the bridge: The new art of structural engineering. Princeton University Press, Princeton.

Braune, A., L. Ekhvaia and K. Quante. 2021. Benchmarks für die Treibhausgasemissionen der Gebäudekonstruktion: Ergebnisse einer Studie mit 50 Gebäuden. Deutsche Gesellschaft für Nachhaltiges Bauen – DGNB e.V., Stuttgart.

Brown, R. 2025. The complete Claude code workflow (to build anything). YouTube, 27 July. Available at: https://www.youtube.com/watch?v=dk97zcYaq_o (accessed 9 August 2025).

Burton, I. 1987. Report on reports: our common future: the World Commission on Environment and Development. Environment: Science and Policy for Sustainable Development 29(5): 25–29. doi:10.1080/00139157.1987.9928891.

Canizares, G. 2024. Taming the Rhinoceros: A brief history of a ubiquitous tool. Perspectives in Architecture and Urbanism 1(2): 100020. doi:10.1016/j.pau.2024.100020.

Carbon Leadership Forum. 2025. C.Scale LCA tool updates. YouTube, 3 February. Available at: https://www.youtube.com/watch?v=eRSob2kwL4Q (accessed 9 August 2025).

Clark, D. 2024. Nvidia posts another record as AI boom continues. The New York Times, 22 May. Available at: https://www.nytimes.com/2024/05/22/technology/nvidia-quarterly-earnings-results.html (accessed 9 August 2025).

Clough, R.W. 1960. The Finite Element Method in Plane Stress Analysis. Proceedings of American Society of Civil Engineers 23: 345–347.

Costello, S. 2023. How 2,400 sensors and machine learning models keep Sydney Harbour Bridge spanning the decades. Computerworld, 11 September. Available at: https://www.computerworld.com/article/1658425/how-2400-sensors-and-machine-learning-models-keep-sydney-harbour-bridge-spanning-the-decades.html (accessed 9 August 2025).

Curth, A., N. Pearl, A. Castro-Salazar, C. Mueller and L. Sass. 2024. 3D printing earth: Local, circular material processing, fabrication methods, and Life Cycle Assessment. Construction and Building Materials 421: 135714. doi:10.1016/j.conbuildmat.2024.135714.

de Botton, A. 2008. The architecture of happiness. Vintage International, New York.

DETAIL. 2013. DETAIL Engineering 3: Bollinger + Grohmann. DETAIL, Munich.

Disney, O., Roupé, M., Johansson, M., Ris, J. and Höglin, P. 2023. Total BIM on the construction site: a dynamic single source of information. Journal of Information Technology in Construction 28. doi:10.36680/j.itcon.2023.027.

Drew, L.J., W.H. Langer and J.S. Sachs. 2002. Environmentalism and Natural Aggregate Mining. Natural Resources Research 11: 19–28. doi:10.1023/A:1014283519471.

Duan, K., C.W.K. Suen and Z. Zou. 2023. MARC: A multi-agent robots control framework for enhancing reinforcement learning in construction tasks. arXiv preprint arXiv:2305.14586. doi:10.48550/arXiv.2305.14586.

Duan, S., Q. Shi and J. Wu. 2022. Multimodal sensors and ML-based data fusion for advanced robots. Advanced Intelligent Systems 4(12): 2200213. doi:10.1002/aisy.202200213.

Fenton, S.K., A. Munteanu, K. De Rycke and L. De Laet. 2024. Embodied greenhouse gas emissions of buildings—Machine learning approach for early stage prediction. Building and Environment 257: 111523. doi:10.1016/j.buildenv.2024.111523.

Gengnagel, C. 2022. UP Transformation einer Großstruktur aus Stahlbeton. Beton-und Stahlbetonbau 117(8): 625–633. doi:10.1002/best.202200059.

Gengnagel, C. and E. Brechenmacher. 2023. Kappe+–Überprüfung einer traditionellen Bauweise auf ihre Leistungsfähigkeit für das Bauen von morgen. Bautechnik 100(1): 45–54. doi:10.1002/bate.202200091.

GeoAI. n.d. Deep learning for point cloud processing. GeoAI. Available at: https://geoai.au/deep-learning-for-point-cloud-processing/ (accessed 9 August 2025).

Grainger, G. 2024. Why 2024 is the tipping point for investing in sustainable buildings. World Economic Forum, 22 January. Available at: https://www.weforum.org/stories/2024/01/sustainable-office-buildings/ (accessed 9 August 2025).

Gudipati, V.K. and E.J. Cha. 2021. Surrogate modeling for structural response prediction of a building class. Structural Safety 89: 102041. doi:10.1016/j.strusafe.2020.102041.

Guo, D., D. Yang, H. Zhang, J. Song, R. Zhang, R. Xu et al. 2025. Deepseek-r1: Incentivizing reasoning capability in LLMs via reinforcement learning. arXiv preprint, arXiv:2501.12948. doi:10.48550/arXiv.2501.12948.

Heisel, F., K. Schlesier and D.E. Hebel. 2019. Prototypology for a circular building industry: the potential of re-used and recycled building materials. IOP Conference Series: Earth and Environmental Science 323(1): 012023. doi:10.1088/1755-1315/323/1/012023.

Henley, J. 2025. Sites without sound: Oslo leads in quiet, low-emission electric construction. The Guardian, 10 January. Available at: https://www.theguardian.com/world/2025/jan/10/oslo-leads-quiet-low-emission-electric-vehicles-building-sites (accessed 9 August 2025).

Hiti, C.J., J. Chang, K. Gwal, E. Escobedo, M. Rea and J. Bindra. 2021. The new normal: Coronavirus pandemic response utilizing Microsoft SharePoint. Journal of Digital Imaging 34(2): 257–262. doi:10.1007/s10278-021-00419-4.

Ho, J., A. Jain and P. Abbeel. 2020. Denoising diffusion probabilistic models. Advances in Neural Information Processing Systems 33: 6840–6851.

IBM. n.d. AI agents vs AI assistants. IBM Think. Available at: https://www.ibm.com/think/topics/ai-agents-vs-ai-assistants (accessed 9 August 2025).

IEA. 2023. World energy outlook 2023. International Energy Agency, Paris.

India Sea Trade News. 2024. Siemens implements automation solutions at port of NEOM. India Sea Trade News, 23 July. Available at: https://indiaseatradenews.com/siemens-implements-automation-solutions-at-port-of-neom/ (accessed 9 August 2025).

IPCC. 2021. Summary for policymakers. In: Climate change 2021: The physical science basis. Contribution of Working Group I to the Sixth Assessment Report of the Intergovernmental Panel on Climate Change. Masson-Delmotte, V., P. Zhai, A. Pirani, S.L. Connors, C. Péan, S. Berger et al. (eds.). Cambridge University Press, Cambridge, United Kingdom and New York, USA, pp. 3–32. doi:10.1017/9781009157896.001.

Ipsos. 2023. AI: making the world more nervous. Ipsos, 10 July. Available at: https://www.ipsos.com/en/ai-making-world-more-nervous (accessed 9 August 2025).

Jumper, J., R. Evans, A. Pritzel, T. Green, M. Figurnov, O. Ronneberger et al. 2021. Highly accurate protein structure prediction with AlphaFold. Nature 596(7873): 583–589. doi:10.1038/s41586-021-03819-2.

Karpathy, A. 2023. nanoGPT. GitHub, Available at: https://github.com/karpathy/nanoGPT (accessed 9 August 2025).

Karpathy, A. 2025a. Deep dive into LLMs like ChatGPT. YouTube, 5 February. Available at: https://www.youtube.com/watch?v=7xTGNNLPyMI (accessed 9 August 2025).

Karpathy, A. 2025b. How I use LLMs. YouTube, 28 February. Available at: https://www.youtube.com/watch?v=EWvNQjAaOHw (accessed 9 August 2025).

Langley, H. 2025. $100 million for an AI genius? It's a bargain, says former Google HR boss. Business Insider, 22 July. Available at: https://www.businessinsider.com/100-million-dollar-ai-genius-bargain-google-hr-boss-2025-7 (accessed 9 August 2025).

Lewis, P., E. Perez, A. Piktus, F. Petroni, V. Karpukhin, N. Goyal et al. 2020. Retrieval-augmented generation for knowledge-intensive NLP tasks. Advances in Neural Information Processing Systems 33: 9459–9474.

Long, E. 2020. How construction robotics contribute to sustainable buildings. The BE Blog: Building Enclosure Online. Available at: https://www.buildingenclosureonline.com/blogs/14-the-be-blog-building-enclosure/post/89389-how-construction-robotics-contribute-to-sustainable-buildings (accessed 9 August 2025).

Ma, Y.J., J. Hejna, C. Fu, D. Shah, J. Liang, Z. Xu et al. 2024. Vision language models are in-context value learners. The Thirteenth International Conference on Learning Representations.

Mahadevan, P. 2019. Sand mafias in India. Disorganized crime in a growing economy. Global Initiative Against Transnational Organized Crime, Geneva, Switzerland.

Mann, J. 2024. Jensen Huang hails Elon Musk and xAI for building an AI supercomputer at 'superhuman' speed. Business Insider, 14 October. Available at: https://www.businessinsider.com/jensen-huang-elon-musk-supercomputer-xai-grok-2024-10 (accessed 8 August 2025).

McKinsey & Company. 2025. Accelerating decarbonization across the farming supply chain. McKinsey & Company, 4 February. Available at: https://www.mckinsey.com/industries/industrials-and-electronics/how-we-help-clients/accelerating-decarbonization-across-the-farming-supply-chain (accessed 9 August 2025).

Meta. 2024. Advancing low carbon concrete in our data centers. Meta Sustainability, 19 December. Available at: https://sustainability.atmeta.com/blog/2024/12/19/advancing-low-carbon-concrete-in-our-data-centers/ (accessed 9 August 2025).

Mikolov, T., I. Sutskever, K. Chen, G.S. Corrado and J. Dean. 2013. Distributed representations of words and phrases and their compositionality. Advances in Neural Information Processing Systems 26.

Mnih, V., K. Kavukcuoglu, D. Silver, A.A. Rusu, J. Veness, M.G. Bellemare et al. 2015. Human-level control through deep reinforcement learning. Nature 518(7540): 529–533. doi:10.1038/nature14236.

Morgan Stanley. 2024. Powering the AI revolution: AI energy demand and opportunity. Morgan Stanley, 8 March. Available at: https://www.morganstanley.com/ideas/ai-energy-demand-infrastructure (accessed 9 August 2025).

Munkhdalai, T., M. Faruqui and S. Gopal. 2024. Leave no context behind: Efficient infinite context transformers with infini-attention. arXiv preprint arXiv:2404.07143, 101. doi:10.48550/arXiv.2404.07143.

MVRDV. 2023a. Innovation Park Artificial Intelligence (IPAI). Available at: https://www.mvrdv.com/projects/952/innovation-park-artificial-intelligence-ipai (accessed 9 August 2025).

MVRDV. 2023b. NEXT climate carbon guidelines poster A2. MVRDV, Rotterdam. Available at: https://mvrdv.com/media/uploads/NEXT%20Climate%20Carbon%20Guidelines%20Poster%20A2.pdf (accessed 9 August 2025).

NVIDIA. n.d. Vision language models. NVIDIA. Available at: https://www.nvidia.com/en-us/glossary/vision-language-models (accessed 8 August 2025).

Ohnsman, A. 2025. Waymo vets are automating construction sites with self-driving dirt diggers. Forbes, 16 July. Available at: https://www.forbes.com/sites/alanohnsman/2025/07/16/waymo-vets-are-automating-construction-sites-with-self-driving-dirt-diggers/ (accessed 8 August 2025).

OpenAI. 2024. Learning to reason with LLMs. OpenAI, 12 September. Available at: https://openai.com/index/learning-to-reason-with-llms (accessed 9 August 2025).

OpenAI. 2025a. Addendum to GPT-4o system card: Native image generation. 25 March. Available at: https://openai.com/index/gpt-4o-image-generation-system-card-addendum/ (accessed 9 August 2025).

OpenAI. 2025b. Introducing operator. OpenAI, 6 February. Available at: https://openai.com/index/introducing-operator/ (accessed 9 August 2025).

Ouyang, L., J. Wu, X. Jiang, D. Almeida, C. Wainwright, P. Mishkin et al. 2022. Training language models to follow instructions with human feedback. Advances in Neural Information Processing Systems 35: 27730–27744.

Preisinger, C. 2013. Linking structure and parametric geometry. Architectural Design 83(2): 110–113. doi:10.1002/ad.1564.

Robles y Zepf, C. and P. Schaefer. 2024. Sustainable data centers—the German Energy Efficiency Act: What data center operators need to consider now and in the future. Mayer Brown, 19 February. Available at: https://www.mayerbrown.com/en/insights/publications/2024/02/sustainable-data-centers-the-german-energy-efficiency-act-what-data-center-operators-need-to-consider-now-and-in-the-future (accessed 9 August 2025).

Röck, M. and A. Sørensen. 2022. mroeck/Embodied-Carbon-of-European-Buildings-Database: v1.0.1 – Official re-release via Zenodo (v1.0.1_Zenodo). Zenodo. doi:10.5281/zenodo.6671558.

Rombach, R., A. Blattmann, D. Lorenz, P. Esser and B. Ommer. 2022. High-resolution image synthesis with latent diffusion models. Proceedings of the IEEE/CVF Conference on Computer Vision and Pattern Recognition (CVPR): 10684–10695.

Roose, K. 2025. 'Vibe coding' and the future of AI software programming. The New York Times, 27 February. Available at: https://www.nytimes.com/2025/02/27/technology/personaltech/vibecoding-ai-software-programming.html (accessed 9 August 2025).

Roudman, S. 2025. The debate on the Bank of America Tower continues. Metropolis Magazine, 12 March. Available at: https://metropolismag.com/viewpoints/the-debate-on-the-bank-of-america-tower-continues/ (accessed 9 August 2025).

Ruiz López, A., D.M.G. Taborda, A. Tsiampousi, A.M.G. Pedro and S. Hardy. 2024. Applying machine learning to the development of surrogate models for shafts in clay. Proceedings of the XVIII European Conference on Soil Mechanics and Geotechnical Engineering, ed. Nuno Guerra, CRC Press, pp. 795–798. doi:10.1201/9781003431749-133.

Sandalow, D., C. McCormick, A. Kucukelbir, J. Friedmann, M. Nachmany, H. Lee et al. 2024. Artificial Intelligence for Climate Change Mitigation Roadmap (Second Edition). ICEF Innovation Roadmap Project. doi:10.7916/2j4p-nw61.

Schiferli, F. and O. Winkler. 2022. Kulturgebäude "Buitenplaats Brienenoord" in Rotterdam. Deutsche Bauzeitung, September. Available at: https://www.db-bauzeitung.de/architektur/kulturbauten/ (accessed 9 August 2025).

Schlaich Bergermann Partner. 2025. Cooling Tower Schmehausen. Available at: https://www.sbp.de/en/project/cooling-tower-schmehausen (accessed 9 August 2025).

Shi, M., F. Liu, S. Wang, S. Liao, S. Radhakrishnan, Y. Zhao et al. 2024. Eagle: Exploring the design space for multimodal LLMs with mixture of encoders. arXiv preprint arXiv:2408.15998. doi:10.48550/arXiv.2408.15998.

Shojaee, P., I. Mirzadeh, K. Alizadeh, M. Horton, S. Bengio and M. Farajtabar. 2025. The illusion of thinking: Understanding the strengths and limitations of reasoning models via the lens of problem complexity. arXiv preprint arXiv:2506.06941. doi:10.48550/arXiv.2506.06941.

Show it better. 2024. How MVRDV is using AI to design their buildings. YouTube, 20 February. Available at: https://www.youtube.com/watch?v=dvKAyTRptkw&t=83s (accessed 9 August 2025).

Silver, D., J. Schrittwieser, K. Simonyan, I. Antonoglou, A. Huang, A. Guez et al. 2017. Mastering the game of Go without human knowledge. Nature 550(7676): 354–359. doi:10.1038/nature24270.

Silver, D., T. Hubert, J. Schrittwieser, I. Antonoglou, M. Lai, A. Guez et al. 2018. A general reinforcement learning algorithm that masters chess, shogi, and Go through self-play. Science 362(6419): 1140–1144. doi:10.1126/science.aar6404.

United Nations Environment Programme and Global Alliance for Buildings and Construction. 2024. Global Status Report for Buildings and Construction – Beyond foundations: Mainstreaming sustainable solutions to cut emissions from the buildings sector. https://wedocs.unep.org/20.500.11822/45095.

United Nations Environment Programme and Yale Center for Ecosystems + Architecture. 2023. Building materials and the climate: Constructing a new future. United Nations Environment Programme, Nairobi.

van der Lugt, P. 2012. Carbon storage utilising timber products. Environment Industry Magazine 2012(6): 76–80.

Vaswani, A., Shazeer, N., Parmar, N., Uszkoreit, J., Jones, L., Gomez, A.N. et al. 2017. Attention is all you need. Advances in Neural Information Processing Systems 30: 5998–6008.

Vincent, J. 2022. An AI-generated artwork's state fair victory fuels arguments over 'what art is'. The Verge, 1 September. Available at: https://www.theverge.com/2022/9/1/23332684/ai-generated-artwork-wins-state-fair-competition-colorado (accessed 9 August 2025).

Watson, N. 2020. Lean design: 10 things to do now. The Structural Engineer: journal of the Institution of Structural Engineer 98(8): 12–14.

Wevolver. 2024. Edge AI market analysis and trends. In: 2024 state of edge AI report, Ch. 1. Wevolver. Available at: https://www.wevolver.com/article/2024-state-of-edge-ai-report/edge-ai-market-analysis-and-trends (accessed 9 August 2025).

Wikipedia. 2025. Steglitzer Kreisel. 12 March. Available at: https://en.wikipedia.org/wiki/Steglitzer_Kreisel (accessed 9 August 2025).

Williams, C.J.K. 2001. The analytic and numerical definition of the geometry of the British Museum Great Court roof. In: Mathematics & design 2001. Deakin University, Geelong, Australia, pp. 434–440.

World Business Council for Sustainable Development and Arup. 2021. Net-zero buildings: Where do we stand? Arup, 8 July. Available at: https://www.wbcsd.org/resources/net-zero-buildings-where-do-we-stand/ (accessed 9 August 2025).

Ye, H., J. Zhang, S. Liu, X. Han and W. Yang. 2023. IP-Adapter: Text Compatible Image Prompt Adapter for Text-to-Image Diffusion Models. arXiv. doi:10.48550/arXiv.2308.06721.

Zhang, G.-Q., B. Wang, J. Li and Y.-L. Xu. 2022. The application of deep learning in bridge health monitoring: a literature review. Advances in Bridge Engineering 3(1): 22. doi:10.1186/s43251-022-00078-7.

Zhang, L., A. Rao and M. Agrawala. 2023. Adding conditional control to text-to-image diffusion models. Proceedings of the IEEE/CVF International Conference on Computer Vision: 3836–3847.

Zhang, L., T. Lu, F. Wang and Y. Xia. 2025. Over 25-year monitoring of the Tsing Ma suspension bridge in Hong Kong. Journal of Civil Structural Health Monitoring 15(1): 263–283. doi:10.1007/s13349-024-00842-5.

Zhang, X., H. Chen, J. Sun and X. Zhang. 2024. Predictive models of embodied carbon emissions in building design phases: Machine learning approaches based on residential buildings in China. Building and Environment 258: 111595. doi:10.1016/j.buildenv.2024.111595.

Zhou, S. and X. Ye. 2023. Approximating high-dimensional minimal surfaces with physics-informed neural networks. arXiv preprint, arXiv:2309.02589. doi:10.48550/arXiv.2309.02589.

Zhu, A., T. Dai, G. Xu, P. Pauwels, B. De Vries and M. Fang. 2023. Deep reinforcement learning for real-time assembly planning in robot-based prefabricated construction. IEEE Transactions on Automation Science and Engineering 20(3): 1515–1526. doi:10.1109/TASE.2023.3236805.

ZillizTech. 2025. deep-searcher. GitHub repository. Available at: https://github.com/zilliztech/deep-searcher (accessed 9 August 2025).

Zitkovich, B., T. Yu, S. Xu, P. Xu, T. Xiao, F. Xia et al. 2023. RT-2: Vision-Language-Action Models Transfer Web Knowledge to Robotic Control. In Proceedings of The 7th Conference on Robot Learning: 2165–2183. PMLR.

Index

For Product Safety Concerns and Information please contact our EU representative GPSR@taylorandfrancis.com Taylor & Francis Verlag GmbH, Kaufingerstraße 24, 80331 München, Germany